Morphogenesis of Meaning

European Semiotics / Sémiotique Européenne

edited by / edité par
Per Aage Brandt and/et Wolfgang Wildgen

Volume 3

PETER LANG
Bern · Berlin · Bruxelles · Frankfurt a.M. · New York · Wien

Jean Petitot
petitot@poly.polytechnique.fr

Morphogenesis of Meaning

Translated by Franson Manjali
Jawaharlal Nehru University. fmanjali@tatanova.com

PETER LANG
Bern · Berlin · Bruxelles · Frankfurt a.M. · New York · Wien

Bibliographic information published by Die Deutsche Bibliothek
Die Deutsche Bibliothek lists this publication in the Deutsche Nationalbibliografie; detailed bibliographic data is available on the Internet at ‹http://dnb.ddb.de›.

British Library and Library of Congress Cataloguing-in-Publication Data:
A catalogue record for this book is available from *The British Library*, Great Britain, and from *The Library of Congress*, USA

ISSN 1423-5587
ISBN 3-03910-104-8
US-ISBN 0-8204-6858-4

Hochfeldstrasse 32, Postfach 746, CH-3000 Bern 9
info@peterlang.com, www.peterlang.com, www.peterlang.net

To the memory of my masters
René Thom
and
Algirdas Julien Greimas

Table of Contents

Translator's Introduction

The present work was originally written as part of Jean Petitot's *Thèse d'État* defended in 1982. It was published in 1985 by the Presses Universitaires de France, Paris, in their series *Formes Sémiotiques* under the title *Morphogenèse du Sens. Pour un schématisme de la structure.* The second part of *Morphogenèse du Sens* was published in 1992 by the CNRS Éditions, Paris, under the title *Physique du Sens.*

The importance of Petitot's original French publication can be emphasized on two counts. Firstly, it provided a deep philosophical elaboration of René Thom's Catastrophe Theory (CT) proposed in the mid-seventies. In his preface to *Morphogenèse du Sens,* Thom acknowledged that the theory which had generated great hopes within the scientific community at the time of its launching ended up being merely 'a set of recipes for modeling', or a tool-kit for applied mathematics. He noted that thanks to Jean Petitot's work

> the philosophical project underlying the whole enterprise [...] has been specified, clarified, amplified, and above all restored to its rightful place within the grand philosophical and methodological tradition of the sciences, particularly the social sciences.

Secondly, and as for the work's significance in the social sciences, Thom approvingly refered to Petitot's catastrophist reworking of Jakobson's structural phonology via the notion of 'categorical perception', to his catastrophist modeling of the 'localist' interpretation of the case category which functions as a sort of fulcrum between syntax and semantics (the localist idea has a history beginning from the Byzantines, Maxime Planude and Theodore of Gaza, to Charles Fillmore and John Anderson, via Louis Hjelmslev), and most importantly to his catastrophist schematization of Greimasian theory of semio-narrative structures.

Morphogenèse du Sens was a seminal work which exerted a deep influence on the different semio-linguistic schools: Greimas' and Coquet's French schools, Eco's Italian school in Bologna, Urbino, and San Marino, Brandt's Danish school in Aarhus, Wildgen's German school in Bremen, and also Canadian schools in Montreal (Pierre

Ouellet and Pierre Boudon) and Québec (Gilles Ritchot and Gaëtan Desmarais). It has become a key reference and we think it is therefore a good thing to provide its English translation.

The present English version is strictly targeted to a 'scientific' readership. As Jean Petitot says in his *Foreword*, the 'continental' philosophical digressions have been deliberately eliminated almost fully. In the process, those aspects of the book that had made it appear epoch-making in the mid- and late eighties may be found wanting in the English version, but the focusing on its scientific 'hard-core' may be more attractive and advantageous, especially to those who are familiar with the dynamical modeling perspectives that have emerged in large numbers in the cognitive sciences in general during the nineties, even if its impact is yet to be felt on the generally slow-moving intellectual horizon of Linguistics (and Semiotics) in a clearly discernible form.

Jean Petitot told me he would prefer to see this pioneering work as a sort of 'retrospective contribution' to the ongoing trends in dynamical modeling, or as a kind of reminder of a strong antecedent which was relatively original for the Anglo-American academic world during its period of euphoria with the Chomsky-Fodor type of formalist cognitivism as well as with other forms of logicism, and also as something that is capable of providing certain fresh insights into the relatively new dynamical paradigm which has blossomed under the aegis of the 'connectionist' research enterprise in cognitive science.

The central issue dealt with in this book is that of *structure*. More precisely, with the question of assigning a *physical and dynamical* basis to structure in linguistics and semiotics. The classical problem with structure has always been to conciliate its formal essence with its phenomenal filling-in, its discrete 'form' with its continuous 'matter' (to use Hjelmslevian terms). For instance, the categories of linguistic structuralism, beginning with the phoneme, etc., are not conceived classically as natural categories, but merely as conceptual ones which are projected onto the real world. The methodological strategy employed in this regard is to suggest that the structural unit, irrespective of where it occurs, and particularly the phonemic unit, is a type subsuming one or more natural tokens, e.g., the phones. In a phonemic analysis, the differences between the phones are identified as distinctive or not. Once the phonetic/phonemic differences are identified, and the distinctive (phonemic) units established, the latter are arranged in paradigms, and are seen as being available for combinatory (syntagmatic) deployment. But the point that is missed in this classical formalist perspective is that there are no abstract categories in nature; categories are largely mental products resulting from a *process* of discretely dividing up the natural entities. These natural entities do not exist as such as *discrete* entities, but form part of a *continuous* substratum. Therefore, a formalization of the structural categories that exist merely as constructs is handicapped by the fact that it leaves behind the continuous and the natural substratum from which structure and its categories inevitably emerge.

Hence the importance of using dynamical models which can explain how *qualitative discontinuities* can emerge from the organization of the continuum, in such a way that it can be categorized and discretized. Catastrophist models yielded the first examples of such algorithms generating discontinuities. Using them, Petitot interpreted the qualitative and the privative oppositions that form the basis of Jakobson's (phonological) distinctive feature analysis in terms of the catastrophes of conflict and bifurcation respectively.

The question of syntactic structure presents an even more interesting picture. Chomskyan axiomatics did go beyond the earlier 'item and arrangement' approach in this domain, by introducing a principle of generativity, essentially based in two sets of rules, those of recursivity and transformation. The apparent autonomy of Chomsky's generative device in fact masks the rootedness of the syntax of natural languages in the structures of action and perception, in other words, the partial analogy that exists between the structure of language and the structure of the experienced external world.

An investigation of the core grammatical structure of natural language reveals not so much an infinite generativity of sentence structures as Chomsky had once claimed, but rather an auto-limitation imposed by the patterns of action in the external world and its perceptual reception by the language-user. Thom's first important intervention in linguistic theory was to question the 'pure and simple idolatory' of the formalist notion of generativity, and to insist on the need for explaining the auto-limitation of the generative capacity itself.

It is here that a grammatical analysis must turn to some of the non-formalist (and rather realist and semanticist) perspectives on the case-structures. More specifically, the actantial perspective of Lucien Tesnière and the localist theory adopted by Hjelmslev, Anderson and Fillmore become relevant. Tesnière had, in his 'stemmatic' analysis of the sentential syntax, granted centrality to the verbal node and regarded the noun phrases as 'actants' that are 'dependent' on that central node. Sentence-meaning was understood, not as the resultant of a combinatorics of word-meanings, but as something configurationally available in a gestalt-like manner. It was composed holistically with the verb conveying the action part of the sentence, and the 'actants' playing the role of participants in the action. Tesnière was explicit about the theatrical imagery (in fact he refers to 'a little drama') while speaking of sentence-structure and its meaning. It is of interest for us to note here that such a view of the sentence and its meaning was precisely what was proposed by the early Indian grammarians in whose verb-centered analysis the term *karaka* is an exact equivalent of the Tesnierian 'actant'. And moreover, for Bhartrhari, comprehension of sentence-meaning is equated with a gestalt-like perception, or *citra-jñaana* (pictorial knowledge).

Hjlemslev too, pursuing his project of a pure structuralism arrived at a perspective not too distant from the above one. In his celebrated book *La catégorie des cas,* after presenting a historical survey of various views on the case-category, he concludes that case cannot be a logical category, but only a structural one. He fully embraces the localist hypothesis of the cases coming down to us from the Byzantine scholars referred to above via the 19th century Kantian linguist Wüllner. In the final analysis, the case is for Helmslev, a category that signifies spatial relations between two objects. He defines these relations along three 'dimensions,' namely, Direction (Distancing and Nearing), Subjectivity-Objectivity and Coherence (with or without contact).

Thom has applied CT to define the genesis of the grammatical (case) structures from the actantial dynamics (derived from Tesnière) on a spatial substratum. The set of seven elementary catastrophes functions as the founding principle for the deduction of the grammatical cases. As a characteristic example, Thom gives the following schemata for the Accusative case (or, the 'actantial graph' for capture):

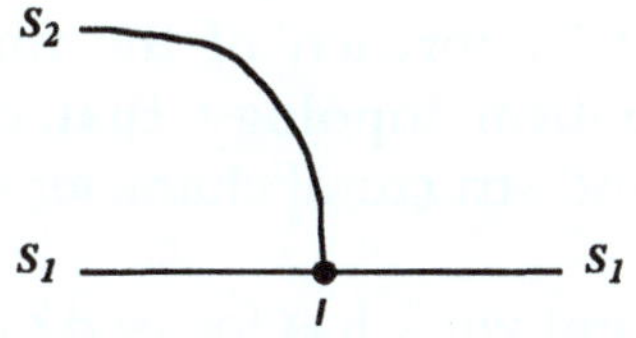

where S_1 and S_2 stand for the paths, in time, of the actants, and I the point of intersection where the sudden disappearance of S_2 takes place.

The above actantial graph is just one of a list of 18 'archetypal morphologies' that Thom has proposed, which are derived from the set of elementary catastrophes. These archetypal morphologies show more finely the correspondence between the topological graphs and the case structures. Thom's topologico-dynamical analysis of syntax-semantics thus involves a synthesis of the actantial syntax, the case grammar and the idea of morphogenesis coming from CT. The main philosophical import of Thom's theory is that it retains an essential continuity between the physical and the phenomenological modes of existence, something that the logicist approaches do not wish to do or are incapable of doing. In the present case, the appearance of phenomenological difference is preceded by a physical process of differentiation of an initially continuous state to yield discrete entities.

Thus CT allows to deduce the qualitatively differentiated case-structures from a topologico-dynamic physical substratum. It provides a principle of identifying and categorizing the finite set of core grammatical (case) structures which in the natural world appear as infinitely varied occurrences of physical or physically-based actions. The main merit of Petitot's work in this regard lies not in proposing the original intuition of the connection between the CT and case theory, but in meticulously establishing the place and the relevance of CT as a viable dynamical approach (what Petitot will rename as a "morphodynamical" approach), in contrast to the various formalist approaches, within contemporary linguistic theory. Petitot's subsequent researches have established contacts with the dynamical approaches in linguistics present explicitly or implicitly in the works of Per Aage Brandt, Leonard Talmy, Ronald Langacker, and George Lakoff. But then linguistics is still to wake up to the fact that 'mathematical linguistics' based on a logico-algebraic formalization which was fashionable during the fifties and sixties has virtually given in to the 'morpho-dynamical' approach of the nineties that employs a sophisticated mathematical topology that can better handle the inherently dynamical and structural character of the core grammar of natural language.

Petitot's more recent work has focused on dynamic modeling in visual perception. An excellent paper which presents Petitot's perspectives on grammar and visual perception is 'Morphodynamics and Attractor Syntax: Constituency in Visual Perception and Cognitive Grammar' that has appeared in *Mind as Motion – Explorations in the Dynamics of Cognition* edited by Robert F. Port and Timothy van Gelder (MIT Press, 1995). It gives a comprehensive picture of Petitot's morphodynamical approach which is now very much part of the dynamical tradition of doing connectionist cognitive science, on either side of the Atlantic. Petitot would like to see it as a sort of synthesis between on the one hand the European theoretical traditions of gestalt theory and structuralism, and on the other the American traditions of Cognitive Linguistics and the dynamical mathematical modeling towards a connectionist AI.

With regard to the analysis of semio-narrative structures (which comes down from V. Propp to A.-J. Greimas via C. Lévi-Strauss), Jean Petitot's attempt has been to theoretically develop the inherent topological potential of the semiotic square by applying CT. This he does by providing a schematization of Greimas' structures of elementary signification and a catastrophist interpretation of the latter's actantial model of narrative structure. Applying the theory on Greimas' model, Petitot suggests that the relations associated with the qualitative and privative oppositions of the semiotic square could be schematized by means of the catastrophe of Conflict of minimal complexity and that of Bifurcation of minimal complexity respectively. This shift, he thinks, is in tune with the topological potential of the square, and involves the abandonment of a logico-combinatory method which is not suitable for a method which must explain the emergence of the structure from a physical substratum. The main merit claimed for the catastrophist model in narrative semiotics is that it can schematize the 'undefinable concepts' of the previous formalist framework. The 'morphogenesis' of the square can be modeled as a 'procession' of elementary catastrophes. At a more complex level, the entire 'canonical formula' of narrative structures as proposed by Lévi-Strauss can be understood in terms of the schemas for two *coupled* qualitative oppositions, represented by a 'double cusp' (which is an intricated singularity).

The 'conversion' that gives rise to the Greimasian actantial model from the syntactic operations on the content values is seen in terms of the actantial graphs associated with the elementary catastrophes. For example, Petitot shows that the conversion $S \cup O \rightarrow S \cap O$ (i.e., a state of disjunction between the Subject and the Object-of-value becoming a state of conjunction between the Subject and the Object-of-value) can be described by means of the actantial graph of 'capture'. As regards the intentional and/or metapsychological dimension which defines the Subject-Object relationship of the interaction, Petitot reminds us that Thom's archetypal morphologies are indeed actantial schemas deeply rooted in the behavioral structures of living beings.

Let me conclude with a personal note. This translation has taken a long period of gestation. It began as something of a hobby during a stay at Maison de l'Allemagne, Cité Universitaire, Paris, while pursuing post-doctoral studies in Linguistics at the Sorbonne. Subsequently, it became a very serious endeavour, with constant encouragement from Jean Petitot. But eventually, we let it grow from being a mere translation into a revised – and even a 'revisited' – version.

I must acknowledge sources of material support this translation project has received at various points during the last ten years: Maison des Sciences de l'Homme, Paris, Indian Council of Social Science Research, New Delhi, and CNRS, Paris. I would like to particularly thank Monsieur Maurice Aymard, Administrator of the Maison des Sciences de l'Homme for the faith he posed in me. The most concerted collaborative effort went into the making of this version of the book during my stay in 1997 at Maison Suger situated in the throbbing heart of Paris. It finally looked like nearing completion during Jean Petitot's visit to the Indian Institute of Advanced Study, Shimla. It has been a great pleasure working with him.

Franson Manjali
Indian Institute of Advanced Study
Shimla

Preface to the English edition

This English version of 'Morphogenèse du Sens' looks more like a 'revisited' edition than just a simple translation. Indeed, Franson Manjali not only did a remarkable job but, due to his deep competence in cognitive linguistics, as can be evidenced from his book *Nuclear Semantics* (Bahri, 1991), he made many important suggestions which enabled me to improve upon the original text. This new version is now metaphysically 'lighter' and more completely focused on its scientific substance. The 'continental' philosophical digressions have been almost completely expunged.

Supposing this book can have any relevance, I think it is mainly as a precursor of the works on topological and dynamical models which have become so widely accepted in the cognitive sciences during the nineties. Thirty years ago, the very idea that physico-mathematical models of this type could be developed for explaining perceptual, linguistic, and semiotic structures was not clearly understood. It was taken for granted that the only available formalization in the cognitive science fields had to be, for principled reasons, of a logico-algebraic and combinatorial type. In this context, René Thom's seminal idea of an alternative morphodynamical paradigm triggered off a true scientific revolution. It settled the basis for a dynamical approach to higher level cognitive tasks such as categorization and syntax.

As far as I know, it was Christopher Zeeman who introduced the first dynamical approach for explaining the links between neuroscience and psychology. In his seminal 1965 article *Topology of the Brain*, he introduced the key idea that brain activity must be modeled by dynamical systems on high dimensional configuration spaces of neural activities. Mental states were then identified with *attractors* of these dynamics, their content with the topological structure of the attractors, and the flow of consciousness with a 'slow' temporal evolution of the neural dynamics. Consequently, the strategy for explaining mental phenomena was to use the mathematical theory of dynamical systems (global analysis) – especially theorems concerning the general structure of the attractors and their bifurcations – for drawing empirical conclusions from this dynamical perspective.

This strategy was very clearly outlined by Zeeman in his 1976 article, 'Brain modelling' :

> What is needed for the brain is a medium-scale theory. [...] The small-scale theory is neurology : the static structure is described by the histology of neurons and synapses, etc., and the dynamic behaviour is concerned with the electrochemical activity of the nerve impulse, etc. Meanwhile the large-scale theory is psychology : the static structure is described by instinct and memory, and the dynamic behaviour is concerned with thinking, feeling, observing, experiencing, responding, remembering, deciding, acting, etc. It is difficult to bridge the gap between large and small without some medium-scale link. Of course the static structure of the medium-scale is fairly well understood, and is described by the anatomy of the main organs and main pathways in the brain. [...] But what is strikingly absent is any well developed theory of the dynamic behaviour of the medium-scale. Question : what type of mathematics therefore should we use to describe the medium-scale dynamic? Answer : the most obvious feature of the brain is its oscillatory nature, and so the most obvious tool to use is differential dynamical systems. In other words for each organ *O* in the brain we model the states of *O* by some very high dimensional manifold *M* and model the activity of *O* by a dynamic on *M* (that is a vector field or flow on *M*). Moreover since the brain contains several hierarchies of strongly connected organs, we should expect to have to use several hierarchies of strongly coupled dynamics. Such a model must necessarily remain implicit because it is much too large to measure, compute, or even describe quantitatively. Nevertheless such models are amenable in one important aspect, namely their discontinuities. (Zeeman, 1977: 287)

It is precisely using these results of global analysis, bifurcation theory and singularity theory, that René Thom worked out his research program leading from physics to cognitive sciences, including linguistics. His main idea was to use these tools for developing a unified mathematical theory of natural morphologies and cognitive structures.

He showed, first of all, that, insofar as it concerns the system of relations which links up parts within a whole, every structure is reducible to a (self)-organized and (self)-regulated morphology. But, as we will see in a detailed manner in this book, every morphology is itself reducible to a system of qualitative discontinuities emerging from the underlying substrate (be it physical, neural, purely geometrical, or even 'semantic'). The theoretical problem was therefore to build up dynamical mechanisms which were able to generate, in a structurally stable way, these discontinuities both at the *local* and the *global* levels.

Deep mathematical theorems have made possible a revolutionary strategy which can be called *dynamical functionalism*. Instead of first defining the generating dynamics *explicitly* and then deriving from it the observable discontinuities, one first describes the observable discontinuities geometrically and then derives from them a *minimally complex* generating dynamics. This minimal explicit dynamics must be conceived of as a simplification of the real implicit generating dynamics.

This dynamical functionalism is not of a classical (e.g. Fodorian) type. Indeed, classical functionalism entails a strict separation between the cognitive and physical levels, the relation between the two being a matter of mere compilation and implementation. This is no longer the case in an emergentist (supervenient) approach. But dynamical functionalism is nevertheless a 'true' functionalism in the sense that emergent structures share properties of universality which are to a large extent independent of the specific physical properties of their underlying substrata.

Such an explanatory paradigm has been extensively developed during the seventies and the early eighties. In physics, and particularly in macrophysics, morphodynamics has innumerable applications. They concern the mathematical analysis of the singularities and discontinuities which emerge at the macro level from underlying micro-physical mechanisms. Here is a very incomplete list : caustics in optics; phase transitions, symmetry breaking and critical phenomena; elastic buckling; defaults in ordered media; shock waves; singularities of variational problems; dissipative structures; changes of regimes in hydrodynamics, routes towards turbulence; deterministic chaos; etc. The main import of these mathematical models is to explain how the observable morphologies which dominate the phenomenologically experienced world can emerge from the underlying physics. They bridge the gap between physical objectivity and common-sense realism, a gap which arose in the aftermath of the Galilean revolution. In that sense, morphodynamics can be considered as the pure mathematical way leading to *qualitative physics*. More than ten years before the computational (Artificial Intelligence) approach was introduced, it showed that the informationally relevant and salient features of macro-physical processes are constituted by their singularities, their qualitative discontinuities and their critical behavior.

But one of the most significant achievements of Thom's paradigm concerned its application to cognitive processes such as perception, action and language. It gave an extraordinary new impulse to traditions such as Gestalt theory, phenomenology and structuralism. It was for the first time that, in cognitive and linguistic matters, *differential geometry could substitute formal logic* as the main mathematical tool.

But Thom and Zeeman proceeded as mathematicians, not in a 'bottom-up' manner, from empirical data first to specific models and, at the end of the line, to theoretical principles, but rather in a 'top-down' manner, from fundamental principles and mathematical structures to empirical data. The advantage of such a strategy was that their perspective was theoretically very well grounded and mathematically very strong. Their dynamical functionalism introduced a new level of functional architecture which could operate as a condition of possibility for the implementation of syntactic processes into the brain dynamics. The limits of such an approach were of course the lack of an effective computational theory to undergird it.

Since the early nineties things have radically changed essentially because dynamical models such as connectionist ones became computationally effective. One can now say along with Tim van Gelder, that the dynamical paradigm has become dominant relative to the logico-combinatorial one. I think that one of the main challenges of future research will be to synthesize the two paradigms.[1]

So, basically this book can be read as a pioneering attempt to introduce morphodynamical models in structural linguistics and semiotics.

I would like to once again acknowledge my debt to Franson Manjali. The long discussions with him on this translation have proved to be a theoretically beneficial opportunity.

---

1 Many aspects of the scientific actuality of morphodynamical semio-linguistics can be found in Brandt, 1992 and Wildgen, 1999.

I want also thank my colleagues and friends Per Aage Brandt and Wolfgang Wildgen for having accepted this text in their Peter Lang series.

Jean Petitot
Indian Institute of Advanced Study
Shimla

INTRODUCTION

1. This work is devoted to a study of the applications of Catastrophe theoretical modeling and of the epistemological issues deriving from it. We will be mainly concerned with the fields of structural linguistics and semio-narrative structures. The investigation proceeds at two levels. At the level of modeling we show that the topological and dynamical syntax conceived of by René Thom allows us to tackle and even partially solve some of the main difficulties encountered in structuralism.[1] At the epistemological level, we examine the relevance of *geometric* notions in the language sciences, and conclude that they provide a *schematization* – in the sense of a *geometrization* of the meaning of theoretical concepts – of the theoretical categories of structuralism. We aim therefore at a *constitution* of the structural domain. Even though this constitution is not strictly of a physical order, to the extent it uses mathematics to reconstruct empirical phenomena, it is of a physical type.

2. From a detailed study of the various structural conceptions, we see that, whatever the domain considered, we come up with a primitive concept of structure whose *formal* content has not yet been adequately mathematized.

(i) In the domain of biological organization, we have to understand how the function of parts in relation to a whole depends on their interdependent *positions*. If a structure can exist, it is because parts are determined reciprocally through a *dynamic process* which defines their *positional values*. This is what Geoffroy Saint Hilaire already called the principle of *connection*.

(ii) In the domain of perceptual organization, a similar problem is posed by the existence of *Gestalt structures*.

(iii) In phonology, the phonemes are conceived of as abstract discriminating units which are equivalence classes of allo-

---

1 By 'structuralism' we mean here the tradition founded by de Saussure and further developped by Troubetzkoi, Jakobson, Tesnière, Hjelmslev, Brøndal, Lévi-Strauss and Greimas.

phones. Now, these classes are also defined by an underlying principle of connection. They are obtained from the *categorization* of audio-acoustic substrata, and are positional values within phonetic paradigms.

(iv) In syntax, the primitive structures are constituted of reciprocally determined actantial places.[1] They also provide, though in a somewhat different way, positional values arising from connections. These connections are *semantic,* and not formal relations. They belong to the form of content (in the sense of Hjelmslev). They are independent of lexical features and constrain the grammatical function of the terms they connect. They belong to a conceptual syntax, and not a formal one. Their content is purely positional.

(v) Finally, in the semiotics of narrative, Greimasian theory employs the phonological and the actantial models to explain semantic and syntactic organizations respectively. It thus combines two structuralist conceptions, and considers semantics in a paradigmatic way (like phonology). The main problem is therefore to understand the linkages between them. The key idea is that of a 'conversion' of the semantic paradigms into actantial (syntactic) interactions, what is called in structuralist traditions the 'projection' of the paradigmatic axis onto the syntagmatic one.

In all these domains, structuralist theory depends crucially and ultimately on the *formal content* that must be ascribed to the category of connection, and thus on the mathematization of the concept of positional value. Only such a schematization can rightfully establish a 'physics' of structures. But it depends, in turn, on the invention of a *geometry of position* that can describe and explain the organization, the stability and the closure of elementary structures as well as the constraints imposed on their combinatorics. It depends on the construction of a new kind of general dynamics, of an original *'analysis situs'*, which still remains a tremendous challenge.

---

1 We use here the terms 'actant', 'actantial', 'actantiality' in the sense of Tesnière (1959) and Greimas (1966). These key words of European linguistics concern the semantic roles of case grammars.

Indeed, as Buffon and Kant had observed, such an *analysis situs* 'is totally lacking in our mathematical sciences'. This 'total lack' has, until now, played the role of a sort of blind spot in our vision of rationality. It has been an 'epistemological obstacle' (in the sense of Bachelard) to the constitution of structural objectivity. It has made structuralist theories to keep swaying between psychological reductionism, idealist vitalism and logical formalism, three positions which are not acceptable except dogmatically.[1]

In linguistics, the formalist approach remains [#] dominant. Based on the fallacious evidence, borrowed from logical positivism, that mathematics is a language which provides the most typical example of syntax/semantics relation, it reduces structures to mere formal combinations. Thus, it is forced to discard the concrete dynamical 'organicity' of structures in favour of a system of abstract relations between terms. As the founders of the *Gestalttheorie* had remarked, this involves a reification of connections, which, by attributing to static terms all that in fact belongs to positional values, ignore the dynamical nature of structures. As far as structures are concerned, formalization is opposed to mathematization. Thus, there exists a conflict between the formal treatment of structures and their 'mathematical physics'. The former is associated with a formal logic of terms and relations while the latter refers to a dynamic topology of places and connections.

3. Catastrophe Theory offers the first instance of analysis situs of structures. It removes, at least in principle, the epistemological obstacle which has until now prevented the constitution of the structural objectivity. We intend to show that this theoretical possibility is also a pratical one.

4. We will reserve another work for a detailed elaboration of the catastrophist formalization of semio-narrative structures.[2] In the present work, we shall focus on some important theoretical questions regarding the above indicated issues.

---

1 See, sections I.2 and I.3.

\# In the seventies. (We will sometimes introduce with the symbol # up to date supplementary footnotes).

2 See Petitot 1992.

In the first chapter we will present a problematized panorama of various critical issues of structuralism. We have tried to give the study as much theoretical opening as possible. We shall refer to structural biology, Gestalt theory, phenomenology and transcendental philosophy. This is necessary in order to trace the 'genealogy' of the structuralist idea and to retrieve all its sharpness and amplitude. In Chapter II, we shall discuss in more detail the two basic structuralist conceptions, namely Jakobsonian phonology and structural syntax. This will allow us in Chapter III to revisit the foundations of Greimas' theory of semio-narrative structures.

5. The main part of this 'physics' of meaning had been developed between 1972 and 1976.[1] If we have postponed its exposition till now,[#] it is because we stumbled upon philosophical difficulties concerning the epistemological status of the modeling of structures as *natural* phenomena. As we know, structures have been traditionnally understood in symbolic terms, that is as constituted of formal relations. A significant leap had to be taken to reach the naturalist conception. The main point is the following. In physical sciences, concepts are not only descriptive, but can also be transformed into algorithms for reconstructing the diversity of phenomena. If we take the structures of meaning as natural phenomena in a physicalist sense, we need to transform the structuralist concepts which describe them into algorithms for reconstructing their diversity.

This is how we were convinced that a *'physics of meaning'* has to be founded on a mathematical schematization of categories of structuralism. In order to stress this idea we have called our project a *'schematism of structure'*.[2]

6. Assuming that our work has some interest and some originality, we hope we will be able to convince the reader that far from becoming obsolete, structuralism is on the contrary in the process of becoming a new frontier of science. We now have the possibility of extending the physical rationalism into a structural rationalism,

1 See Petitot 1977b, 1977c, 1979c, 1979d.

\# That is 1983.

2 Our notion of schematism is not exactly that of Kant's transcendental schema. It concerns schematism as a 'construction' procedure for concepts.

mathematically founded, encompassing symbolic and semiotic orders. We now have the possibility, by extending natural ontology, of naturalizing meaning without any longer having to sway between its symbolic reification and its existential experience.

Torre Pellice, August 1983.

CHAPTER I

Problematic Aspects and Key Issues of Structuralism

In this first chapter we shall describe methodically, though not exhaustively, some of the most significant aspects of dynamical structuralism (Sec. 1 and 2). This will lead us to an inquiry into the conditions of possibility of mathematizing structures (Sec. 3.1). As Gilles Deleuze has shown in a essay that we will discuss (Sec. 3.3), the foundations of structuralism are *topological* – and not logical (Sec. 3.2). Until now, the absence of such foundations have been obfuscated by speculative interpretations because of the lack of any adequate geometry (Sec. 4). In conclusion, we shall briefly summarize the principles of Catastrophe Theory (Sec. 5).

1 Understanding 'structure'

Depending on the domain considered, the concept of structure can have quite different contents and epistemological values. In the case of a mechanical device, a construction, or a work of art, we can generally describe the structure in terms of its design. In the case of physico-chemical systems (e.g. crystals, macromolecules, etc.) we can also derive the structure from the interactions between its components. For example, the progress made in molecular biology and in microbiology has resulted in a decisive advance in the comprehension of the structure (stereo-chemical composition) of DNA chains, proteins, enzymes, membranes, etc. There are of course considerable experimental difficulties. Their solution requires highly sophisticated technologies of observation and reconstruction. But, in principle, even if one does not fully confuse the structure with the observed morphology, even if one considers the former as the organizing principle underlying the latter, the *ontological* status of the reconstructed structures is not at all problematical.

On the contrary, in other domains, such as naturalist biology (taxonomic, anatomical, morphogenetic), perception, anthropology or semio-linguistics, one encounters *non-material supervenient* structures, *abstract* forms of organization which are not directly reducible to systems of components in interaction. This simple fact raises considerable theoretical problems to the extent that we cannot any longer, without further inquiry, regard the structures as empirically given phenomena and objects of experience endowed with a predefined ontological status. The very *objectivity* of structures must then be *constituted* as such and that is why, in all these cases, a deeper reflection leads us:

(i) to promote the organizational concept of structure to the level of a fundamental category of scientific thinking,

(ii) to investigate its objective value, and

(iii) to seek ways to mathematize its categorial content.

In naturalist and descriptive biological sciences, as in social sciences, structuralism represents a *rationalist* attitude, emphasizing the role of theory and formalization. Its point of view is opposed to empiricist reductionism as well as historicist evolutionism. The shifts from atomistic psychology to *Gestalttheorie*, from comparative and historical linguistics of the 'neo-grammarians' to Saussurian structural linguistics in Europe, or from 'behaviorist' linguistics to generative or cognitive grammars in the United States, from biographical and socio-psychological literary criticism to structuralist criticism, etc., are trends in the direction of a general philosophy of systems conceived as rule-governed wholes. In this sense, the horizon of structuralism is that of a theoretical description of *formal dependence relations* which 'organically' connect the parts in a whole.

In so far as it is the ideal *form* of the organization of a substance, a structure is *not* a sensible phenomenon. Though it is invisible as such, its substantial realizations and its effects are observable and can be subjected to well-defined experimental procedures. In this sense, every structure is a theoretical object – and not a fact. If we want to avoid naive idealism, we have to constitute it as an object of experience, as a form emerging from the organization of the substrata where it is implemented. Thus, we encounter here a 'foundational aporia', to use René Thom's expression. As Gilles Deleuze claims, a structure is 'real without being actual, ideal without being abstract'; it is a pure 'virtuality of coexistence which pre-exists being'; it is 'embodied' (implemented) in its substratum, but is never actualized as such.[1] The sensible expression of a structure is always a negation of its ideal essence. That is why, as Krzysztof Pomian observes, all structural approaches substitute the initial observed objects such as language, natural forms, etc. with pairs of objects whose ontological statuses are different:

> parole and langue (Saussure), allophones and phonemes (Jakobson, Trubetzkoi), substance and form (Hjelmslev), systems of kinship and elementary structures of kinship (Lévi-Strauss), performance and competence (Chomsky), empirical morphologies and their underlying dynamics (Thom), etc. Each of the first terms of these pairs (which one might call 'realizations') are accessible to sensory experience, or to observation, and their reality consists in this. Each of the second terms, the structures [...] cannot by definition be perceived or observed; we grant them a reality on the basis of a demonstration, more or less rigorous depending upon the case. The relations between realization and structures are variable, but it is always the latter which render the former stable and intelligible. As a result, structures are defined as the sets of rational and interdependent relations, whose reality is demonstrated, whose description is provided by a theory, and which are realized by a visible or observable object whose stability and intelligibility are conditioned by them.[2]

Given such a status – ideal and non phenomenal in the classical sense –, structures are thus ontologically ambiguous. As Umberto Eco asked:

> Is the structure an object, in such as it is structured, or rather the set of relations which structures the object, and can be abstracted from it?[3]

In fact, as *eidos*, a structure is not detachable from the substance where it is actualized.[4] But must we consider it as *given* or as *posited*? In the first case, one will tend to develop an *ontological* (realist)

1 Deleuze, 1973: 313. We will discuss Deleuze's views in Sec. 3.3.
2 Pomian, 1981: 758.
3 Eco, 1968.
4 Cf. ibid.

conception of structures while in the second, an *epistemological* (nominalist) conception.

Currently, the epistemological interpretation of the category of structure is dominant. It reduces structure to an operational concept whose reality is not ontological but only methodological. However, it should be stressed that all the major structuralists (Saussure, Jakobson, Tesnière, Hjelmslev, Piaget, Lévi-Strauss, Chomsky, Greimas, and lastly, Thom) have been or are 'realists', even if they don't engage in a philosophical quarrel.

In fact, from an epistemological, methodological and 'nominalist' perspective, the concept of structure can only be a *descriptive* concept, indeed empirically based, but epiphenomenal and devoid of any objective value of its own. Though operational, it is nothing more than a theoretical construct, an artefact, and cannot by itself be a genuine scientific notion. In particular, it cannot contribute to the mathematization of phenomena. On the other hand, from a 'realist' perspective, it is a concept, though initially problematic, acquiring beyond its empirical validity, an objective value and a constitutive role. Via the schematization of its categorial content, it becomes a source of algorithms for reconstructing specific classes of phenomena.

If we wish to subject structuralism to a systematic historical 'spectral analysis', we must analyze at least the following trends.

(i) The dynamical structuralism of biological origin, which, starting in German philosophy with the *Naturphilosophie* and Goethe's *Morphologie*, has progressed, via Driesch and D'Arcy Thompson, up to Waddington's concepts of 'morphogenetic field' and 'chreode'. This dynamical structuralism is centered on the problem of morphogenesis.

(ii) The phenomenological and gestaltist structuralism which began early this century on the basis of Brentano's works with Stumpf, Meinong, Ehrenfels, Husserl, Köhler, Koffka, Wertheimer, etc.

(iii) The linguistic structuralism resulting from Saussure's 'epistemological breakthrough'. As we already stressed, it has become one of the basic paradigms in social sciences, be it in Phonology with Jakobson, in Anthropology with Lévi-Strauss, in General Linguistics with Tesnière and Benveniste, or in Semiotics with Hjelmslev and Greimas. This structuralism is twofold:

(a) the 'realist' phenomenological structuralism of Jakobson which maintains close relations with dynamical structuralism and Gestalt theory;

(b) the formalist structuralism ('methodological' and 'epistemological') of Hjelmslev, Lévi-Strauss,[1] Chomsky and Greimas who conceive of structures as 'axiomatized' theoretical objects and solve the question of their ontological status by embedding them in genetically determined cognitive capacities.

(iv) The epigenetic and cognitive structuralism of Piaget.

(v) The 'catastrophist' structuralism of René Thom, which is a profound synthesis of the concepts of morphogenesis and structure. It is the first approach to have succeeded in mathematizing structures as theoretical objects.

To get a more complete picture, we must also explain certain general problematics related to the project of structural rationalism. Of these, at least five appear to be essential.

(i) *Experimental methods which provide an access to the structures.* We have seen that structures are ideal and non material, and cannot be directly observed. A first method of access (advocated by Lévi-Strauss) consists in analyzing the *transformations* of structures by variational procedures. Indeed, if a structure identifies itself with a global, internal, and rule-governed system of relations, then every local variation must imply a global transformation manifesting the structure. A second method (that of Chomsky's native speaker conceived of as a language automaton) involves the use of the traditional practice of introspection as part of the experimental procedure.

(ii) *The relation between structure and function.* Ever since the historic debate confronting Geoffroy Saint Hilaire's principle of connection with Cuvier's principle of functional correlation, there has been in biology a dialectical relationship between a physicalist attitude (mechanistic and materialist) endorsing a 'micromerist' reductionist conception, supporting active experi-

1 Lévi-Strauss' conception is more complex. It involves also Jakobsonian and biological structuralisms.

mentation, rooted in physiology and, to day, of essentially neodarwinian inspiration, and a naturalist attitude, endorsing a holistic vitalist conception, supporting common sense observation, based on morphogenesis, and of a somewhat Lamarckian inspiration. But this debate is often a bit skewed, for the phenomena of adaptation (and in particular those of adaptive convergence and co-evolution) show that these two positions are rather complementary, and that it is impossible to privilege one against the other.[1] The real problem is rather to explain the complementarity itself.

(iii) *The relation between structure and teleology (finality).* One of the main reasons for disfavoring the concept of structure since long, has been essentially the fact that, as regards the systematic organization of parts in a whole, it is a teleological idea. To transform it into an operative scientific concept, we must 'definalise' it. This is possible only by way of its mathematization.

(iv) *The formalization of structures.* It has become commonplace to say that the concerted development of general theory of systems, cybernetics, and formal ontology of relationships has enabled the 'axiomatization' of the concept of structure. But it must be emphasized that the mechanistic-formalistic approaches are largely insufficient. As we have seen, they are obtained only by a symbolic reification of structures. They cannot account for the dynamically self-organized and self-regulated emergent (supervenient) forms. In other words, they do not provide an answer to the critical question of the form-substance relationship.

(v) The levels of organization. Structure-function complementarity comes up at all levels of composition and observation. The central question is to define the objective reality of these levels and to understand their correlations.

In this first chapter, we will provide a preliminary sketch of these diverse issues. We will not speak of the 'classical' structuralism which forms part of the comtemporary scientific culture (Saussure's struc-

1 See, Delattre *et al.*, 1973.

turalism, Parsons' structural-functionalism, Harris' and Chomsky's structural linguistics, the structural analysis of economic equilibria, etc.).[1] We prefer rather to focus on:

(i) the still largely unsolved theoretical problems concerning structuralism;

(ii) its morphological, phenomenological, and gestaltist 'accursed' part;

(iii) the 'revolution' represented by the catastrophist turn.

2 Main trends in structuralism: a brief review

In this section we present a brief historical account on some major perspectives on the phenomena of (self-)organization. We are dealing with fields (biology, psychology, phenomenology, anthropology, and semio-linguistics) where the concept of structure is not only a descriptive tool but also a means of going beyond the conflict between the objectivist-reductionist explanations and the idealist-holistic ones (see section 1).

1 For an introduction to structuralism, see for instance the following works : Almansi, 1970; Bach, 1965; Badock, 1975; Barthes, 1966; Bastide, 1962; Benoist, 1975; Benveniste, 1966; R. Boudon, 1968, 1973; P. Boudon, 1981; Broekman, 1974, Cassirer, 1945; Chomsky, 1965; 1966, 1968; Damisch, 1973; Delattre, 1971; Eco, 1963; Ehrmann, 1966; Gandillac et al., 1965; Glucksmann, 1974; Greimas, 1966; Guillaume, 1979; Harris, 1951, 1970; Hawkes, 1977; Hénault, 1979, 1983; Hjelmslev, 1968, 1971; Jacob and Francone, 1970; Jakobson, 1971; Jakobson and Lévi-Strauss, 1962; Katz and Fodor, 1964; Laughlin, 1974; Leach, 1976; Lévi-Strauss, 1949, 1958, 1964–1971; Macksey-Donato, 1970; Maranda, 1966; Marin, 1977; Piaget, 1968; QS, 1973; Raccani and Eco, 1969; Robey, 1973; Saussure, 1915; Sebeok and Osgood, 1965; Segre et al, 1965, Viet, 1965.

2.1 *The aporia of organization in Kant's Critique of the Faculty of Teleological Judgement*

I think we can locate the origin of the modern structural problematic in Kant's treatement of biological organization in terms of *finality* (Kant called it 'the internal finality of natural ends') in his *Critique of the Faculty of Judgement;* more precisely in his demonstration that the theoretical comprehension of organization necessarily required two complementary principles (two 'maxims' of judgement), one reductionist, and the other holistic. Let us briefly trace his arguments.[1]

(i) Given the *a priori* structure of possible experience, we cannot admit of any objective finality in nature. Objectively speaking, nature is necessarily mechanical. In other words, reductionism is the only objectively valid thesis.

(ii) It is however an empirical fact that there exist in nature 'natural ends', i.e., things which are 'cause and effect of themselves',[2] in short, organized living beings. The fundamental features of the natural biological ends are, according to Kant, morphogenesis, regulation (homeostasis), reproduction, and the adaptive relationship with the environment (external finality).

(iii) Now though Kant might have accepted that the progress of physics would, one day, explain mechanistically some of these features, he made the decisive remark that such an explication would still, for *a priori* reasons, be incomplete to the extent that it would not account for the *contingency* of the form of organized beings. For Kant, the contingency of form is part of the 'specific character' of natural ends. Because it eschews the laws of geometry and physics, it can be understood only *reflectively* via the Idea (and not the category) of finality.

(iv) The internal finality is not just organization, but *self*-organization. In a natural end, there exists a reciprocal determination between the parts and the whole. The structure is not that of a mechanism, but the effect of the idea of the whole determining the systematic unity of the form and the connection between

1 See, Petitot, 1982d, for a more detailed account.

2 Kant, 1790: 190.

parts. The organization depends therefore on a 'formative force' (*bildende Kraft*), which not being explicable mechanically, is not objective. That is why it is an 'unfathomable quality', an 'incommensurable abyss' where reductionism, though the only objectively valid maxim, should nevertheless be treated along with the holistic concept of finality.

(v) The reductionist and holistic maxims of judgment seem to be contradictory. They open out therefore to a natural 'dialectic'. But, for Kant, the conflict is not a true antinomy for it concerns only maxims, i.e., prescriptions that a subject must follow for gaining knowledge. Maxims are only heuristics for the comprehension of phenomena. There would be an antinomy only if, moving dogmatically from reflective to determinant judgement, we would use the idea of finality as a constitutive concept, as an objective category. But, even if it is only heuristic, the rational concept of finality is 'as necessary for the *human faculty of judgement* as if it were an objective principle'.[1]

(vi) For Kant, the possibility that a regulative Idea can have the same value as a categorial concept comes essentially from the finite ('discursive', 'non-intuitive') nature of our understanding.

Since Kant, things seem to have notably changed. But this is quite illusory. The epistemological obstacle masterly identified in the *Critique of the Faculty of Teleological Judgement* – namely, the principled impossibility of a physical explanation of the phenomena of morphogenesis, (self-)organization, and regulation – is still far from eliminated. Indeed, the advances in reductionist biology (molecular biology and neo-Darwinism) on the one hand, and in the techniques of cybernetic simulation on the other hand, have given us a lead. But we are still far from understanding how stable and self-regulated structures can emerge from a physico-chemical substratum. The difficulty is not so much experimental as theoretical. What we lack are concepts, not facts. It is only recently that in the physical (non-biological) cases we have been able to explain, using the theory of bifurcations of dynamical systems, how material media can spontaneously self-organize, either purely temporally (oscillating chemical reactions) or, spatio-temporally (spatial patterns of Belousov-

1 Kant, 1790: 218 (our translation).

Zhabotinsky reaction, Bénard's cells, etc.).[1] In this sense, biology still remains, as Jean Piaget had affirmed, 'the key to structuralism'.[2]

2.2 *Structuralism in Biology*

In biology, the structuralist paradigm is a dynamical perspective that appeared whenever the idea of morphogenesis came up. Here, the concept of structure is inseparable from that of form. Therefore it has always been, until recently, tied up with the speculative concept of *entelechy* which goes back to Aristotle. This explains why it was rejected by the anti-Aristotelian reductionists.

The issue began with the *principle of spatial connection* between parts in a whole, introduced by Geoffroy Saint Hilaire, and later taken up by Goethe. In his long and patient meditations on plant morphogenesis, streching from 1770 until his death in 1832, Goethe sought not so much to understand the physico-chemical mechanisms underlying the formation of organisms, as to discover the principle by which an organism is what *it appears* to be.[3] He quickly came to the conclusion that what distinguishes an organism from a machine is the fact that in the case of an organism, the *external* appearance is governed by an *internal* principle producing the *spatial* (external) connections between parts. For Goethe, it was the understanding of this principle which constituted the central theoretical problem in Biology. However, though referring to an empirical phenomenon, the concept of connection is, as we see with Kant, only a 'noumenal' Idea, and not a 'determinant' concept or category. Transgressing the argument of Kant's third *Critique*, Goethe put forward the hypothesis that there existed a *schema* for this Idea, which could share infinite concrete variations. To understand the response of organisms to stimuli as much internal as external, he seeks to determine their constitutive ideal principle, in other words, their *formative laws*.

Goethe gradually recognized this ideal principle *in the spatio-temporal unfolding of an internal organizing force.* According to him, it is

---

1 See, for instance, Prigogine, 1980.

2 Piaget, 1968.

3 For this account of Goethe's conception, see Steiner, 1884. I thank Filomena Molder who introduced me to this remarkable work.

this *'a priori' entelechic* principle that rules the formation of natural ends. But one of the central results of the Kantian Critique is precisely that a noumenal Idea is, in essence, disconnected from space and time. Against Kant, Goethe thought of entelechy as a kind of 'intuitive concept'. Contrary to physics, where concepts are abstractions relative to the sensible world, for him the concept of structure was a real, concrete and perceptual entity. That is why entelechy can be an the intuitive concept and an efficient idea, which by unfolding itself spatio-temporally brings about morphogenesis.

Goethe's answer to the aporia of form in biology was of a speculative nature. It is one of the sources of vitalism. But nevertheless its epistemological value continues to be retained in contemporary trends of dynamical structuralism. As an example we can refer to the defense of structuralism in biology proposed by B. Goodwin and A. Webster, in line with the ideas of the great embryologist Waddington.[1]

Goodwin and Webster present a historical and epistemological analysis of the classical conflict between the structuralist and the neo-Darwinian points of view, the latter being the synthesis of the Darwinian evolutionary theory and molecular genetics. For them, structuralism is opposed to neo-Darwinian empiricism, not at the level of facts, but as a rationalist point of view in which a priori concepts, categories, and principles govern the explanation of empirical data. The central problems they address are those of form and morphogenesis. They investigate the *type* of categoriality necessary to make these concepts intelligible. Now, the main point is that, by its very evidence, the neo-Darwinian paradigm obscures the intelligibility of morphological phenomena. It reduces them to a by-product of evolutionary chance, denying thus any 'laws' of form.

This is essentially due to the fact that this paradigm confuses the concept of *control* with the category of *cause*. The genome controls the form and the development of an organism at the phenotype level. By acting on the genome one can therefore also manipulate its morphological effects. But this causal efficiency does not entail that there are no specific and autonomous constraints for forms. By identifying the genetic control of the phenotype with a determinant cause, the

1 See Webster, Goodwin, 1981; Waddington, 1956, 1957. For the opposite, neo-Darwinian point of view, see, for instance Danchin, 1977.

neo-Darwinian approach assumes that there is nothing to be explained other than the phenomenon of control itself: as Jacques Monod claimed, form is causally reducible to the primary structure of proteins, and all the rest is only a matter of thermodynamical processes of self-organization.

Neo-Darwinism is a materialist reductionism which privileges functional aspects, reduces structural connections and positional organization of parts to a mere spatial contiguity, and subordinates the 'internal finality' to an 'external finality', i.e., to adaptation and selection. It reduces structure to genetics. For it, structure is historically given , and has only an evolutionary necessity as the epigenetic expression of its genetic programme.

Structural rationalism denounces the inconsistency of making history not only the cause of evolution, but also that of *stability* and *invariance* of species.[1] According to its view, an organism is not only a genetically controlled system, but also a structure, that is a totality organized by a system of internal relations satisfying some *'laws' of form*. The realm of organized beings manifests a certain necessity. The structures are neither irreducibly diverse, nor the arbitrary result of evolution.

The fundamental tenet of structural rationalism is that the expression of the genotype into the phenotype cannot be completely understood unless we introduce some sort of *positional information* controlling cellular differentiation. In organized beings there would be a *positional efficiency*, the position selecting metabolic regimes by triggering the right genes. It is the understanding of such positional information and efficiency which constitutes the central theoretical problem of dynamical structuralism.

In the Waddingtonian theory of morphogenetic fields and 'chreodes', the main characteristics of structural organizations are the following:[2]

(i) dynamical genesis, self-regulation and structural stability;

(ii) equipotentiality: structures are not mere systems of interaction of components, but include a reciprocal determination of *places* (positional values);

---

1 On this question, see also Gould, 1977, Gould, Eldrege, 1977.

2 For more details, see for instance, Ruffié, 1982: Chapter XI.

(iii) equifinality and homeorhesis (epigenotype according to Waddington): development is itself structurally stable as a process, and its final state is largelly independent of its initial state;

(iv) the closure of the elementary structures and the existence of constraints, or 'laws' of form;

(v) 'generativity' of forms and the production of complex structures from a closed set of elementary ones.

All these concepts are categories governing morphological phenomena. Their categoriality (which as we shall see later is more 'linguistic' than physical) determines the type of theory we need to render intelligible the morphological and dynamical concept of structure. We see that the main problem is to give them an *objective value*.

2.3 *Gestalt theory and phenomenology*

In psychology, structuralism begins with *Gestalttheorie* where we encounter the same issues, the same problems, the same criticisms of reductionism and the same categoriality that we have already sketched. In his classic introduction (recently republished), Paul Guillaume [1] insists that Gestalt theory is a rationalist monism which introduces the category of structure simultaneously in the physical, the biological and the psychological realms. In psychology, it begins with a criticism of the atomistic view of sensations and of associationism, and maintains close relations with Husserlian phenomenology. The concept of pure sensation is just an experimental artifact, a hypothetical explanatory concept, because a sensation cannot exist without perceptual organization. Indeed peripheral excitations (retinal, for example) are produced by the external stimuli. But they are only local inputs for corresponding global percepts. They are not their determinant cause.

The fundamental hypothesis of Gestalt theory is that it is impossible to reduce perceptions to systems of atomic sensations, since such systems are the product of a construction which involves a *real* transformation of the state of consciousness. Neither the terms nor the relations have an atomic sensorial reality and that is why it is

1 Guillaume, 1979.

necessary to conceive of perceptions as 'complexions' (to use Meinong's term), as Gestalten, i.e., as structures, as

> organic units which are individualized and delimited in the spatial and temporal field of perception or of representation.[1]

These structures, morphologically organized and internally articulated, result from an original formative activity. Their difference with systems of components in interaction again lies in the existence of connections determining positional values. They are *non-compositional* totalities, whose moments do not possess the status of independent parts detachable from the whole.

Instead of attempting a purely phenomenological description of structures like Husserl, or a symbolic-combinatorial description like the formalists, the gestaltists theorized them dynamically as natural biophysical phenomena. To this end, they put forward the hypothesis (masterly confirmed later; see Sec. 3) that 'the principles of dynamics exceed, in their generality, their strictly physical applications.'[2]

As Guillaume emphasized, Gestalt theory views the organized entities, whether physical, biological, or psychological,

> as satisfying very general laws of dynamics pertaining to organized wholes, laws which are neither specifically physical nor psychological, but common to both physics and psychology.[3]

In this regard, Köhler spoke of *Eigenstruktur* governed by a principle of *functional proximity.*

Thus, even before the structuralist trends of the 50's and 60's, as much at the level of natural phenomena as at the level of phenomenology of perception, or of language that mediates between perception and the world, the concept of structure has been deeply reflected upon, early in this century, in Austro-German philosophy where there existed close relations between:

(i) the founders of structural psychology, Stumpf, Meinong and Ehrenfels (all of them students of Brentano; Karl Stumpf taught

---

1 Ibid.: 23.
2 Ibid.: 36.
3 Ibid.: 153.

Husserl and the Gestaltists of the Berlin school, Wertheimer, Köhler and Koffka);

(ii) Husserlian phenomenology;

(iii) Hilbertian axiomatic;

(iv) linguistic reflections of Wittgenstein and the Vienna Circle;

(v) via Jakobson, the linguistic works of the Prague Circle.

Therefore we cannot afford to forget that the roots of modern structuralism are situated at the meeting point of biological naturalism, phenomenology, and Gestalt theory. We have to add a few more words in this regard.[1]

From Brentano on, the classical debate on parts/whole relations has been taken up in a new perspective. Many issues can now be considered as conceptually resolved (for instance, the physical content of causality or interaction, the set theoretic notions of membership and of inclusion of one class into another, the nature of spatial connections in a given space, the relations of syntactic dependence in a logical formula, etc.). But many other issues, crucial for phenomenologists and Gestalt theorists, remain still mostly unresolved.

Some of these are the following.

(i) The problem of the *objective* correlates of the classificatory relation between a genus (higher type) and a species (lower type): what can be the objective validity of classifications, and of a *realist* concept of abstraction?

(ii) That of the objective correlates of the relational 'accidents', either of static type (contractual relations, like kinship relations) or of dynamic type (actantial relations). This central problem (to which we will return in Sec. 2.4) concerns the *states of affairs*, which without being objective in the strict (physical) sense, are nonetheless objective correlates of their linguistic descriptions.

(iii) That of organization, be it biological or perceptual.

---

1 For elaborating these issues, we will refer to the important work of Barry Smith, Kevin Mulligan and their colleagues which throws a fresh light on this tradition. See, Smith, 1982.

(iv) That of *non-detachable* parts in a whole, i.e., that of dependent moments. For example, in the perception of an object, a sensible quality like colour is non-detachable from its spatial extension, other than by abstraction. Similarly, the apparent contour of an object is non-detachable from its extension. It cannot exist independently.

This last problem, namely the relations of dependence between a moment and the whole from which it cannot be detached, has been deeply investigated by Stumpf, Meinong, and Husserl.[1] We can approach it either as a problem of psychology and Gestalt theory, or as a general problem of ontology. This is what Husserl does in conceiving of the relation of dependence as a *formal* concept, and in attempting to 'axiomatize' it in terms of formal ontology. This move is of considerable significance, since it presupposes the realist hypothesis that the relations of dependence (Husserl called them also 'relations of foundation' or 'metaphysical connections') are not only psycho-linguistic but also 'a priori' valid for every field of objects, and therefore possess an objective content. Its consequences are far reaching.

(i) It played a foundational role in Gestalt theory.

(ii) Applied to *syntactic* units, that is, to what Ehrenfels and Meinong called the higher order objects, it strongly influenced the Polish school of logic (particularly, Lesniewski and Ajdukiewicz) and the development of a 'pure logical grammar'.

(iii) It became the theoretical cornerstone of Jakobsonian phonology; the distinctive features are dependent moments 'par excellence'; the phonemes are neither equivalence classes of allophones nor descriptive abstractions, but formal and relational units constituted of dependence relations; they are real relations in the sense of an ontological autonomy of the phonological level (see Sec. 2.5 below).

---

1 For more details, see, Smith, 1982.

2.4 *The states of affairs* (Sachverhalte)

Before taking up the relationship between structuralism and semiolinguistics, let us say a few words on the crucial notion of *Sachverhalt* which relates linguistic structuralism and Gestalt theory, and which relies upon the realist conception of relations of dependence proposed by Husserl. The descriptive relation between language and the external world cannot be reduced to a mere denotative one. To understand it, it is necessary to introduce a *third term*. If we take a sentence describing an external fact (for instance, an actantial interaction), we must suppose that its syntactico-semantic structure possesses an objective correlate, and that there is an 'objective' structuration of the fact – a system of structural connections – which is linguistically expressed. The difficulty is that such a structuration is neither of a physical nor of a linguistic type. It does not have any material existence. It 'subsists' ideally as a morphological articulation of the physical world. It constitutes a third term between expression and reality, which is what is called a state of affairs.[1]

Now, we can consider the concept of state of affairs in two opposite ways:

(i) Either, by equating it with the correlative fact, and ascribing to it only the role of a truth-maker: this is the dominant point of view in the philosophy of language. Whatever be its latter refinements (e.g., intensional logic explaining opaque contexts or the *de dicto/de re* distinction in modal logic), the relation between language and reality still depends on a denotative conception analogous to the relation between syntax and semantics in model-theoretic logic.

(ii) Or, by trying to explain how it can emerge from the external fact as an 'objective' structure, a *phenomenological invariant*, whose reality is neither physical nor symbolic. This second position, much more restrictive than the first, is quite relevant because it shows how linguistic structures are determined by constraints imposed by the structure of reality and of perceptual Gestalts. René Thom holds such a view when he asks:

1 See again Smith, 1982.

> Can't we accept [...] that the factors of phenomenological invariance which create in the observer the sensation of signification, come from the *real* properties of objects of the external world and demonstrate the *objective* presence of formal entities pertaining to these, entities which could be called *'bearers of signification'*.[1]

Obviously, such a claim is acceptable only if we can integrate phenomenological appearance within objective reality and provide a mathematical definition of these 'formal entities' as 'factors of phenomenological invariance'.

As we will see, Catastrophe theory provides the first synthesis between phenomenology and physical objectivity. According to Thom,

> strictly geometrico-topological [morphological] analysis [...] allows us to associate with every spatio-temporal process certain invariants of combinatorial nature [catastrophes] [...] which, by virtue of their fundamental character, can reasonably be thought to play an essential role in the verbal description of the process. Such is the origin, I think, of the original schematism that underlies the linguistic organization of our vision of the world.[2]

> Since the primordial function of language is to transcribe the phenomenological catastrophes of the external world in a form communicable by our organs, [...] the message bearing an autonomous signification inherits the structure of the external catastrophe that it intends to signify.[3]

Lacking such a theoretical device, we might have to conclude with Husserl and Wittgenstein that physics, however perfect it may be, will not yield a description of the phenomenological states of affairs and that the latter are apprehendable only via their linguistic expressions. But then, the two become indistinguishable and we are thus constrained to postulate that a linguistic statement refers to a non-linguistic state of affairs without being able of saying anything of it except in a tautological manner.

This vicious circle, well pointed out in Wittgenstein's *Tractatus*, pervades contemporary formal linguistics as well as purely logicist versions of analytic philosophy. Without a synthesis between phe-

---

1 Thom, 1980a: 170.
2 Thom, 1980c: 24.
3 Thom, 1972a: 329.

nomenology and physical objectivity we cannot escape the dilemma excellently formulated by Pierre Ouellet:

> Is language something which gives entities their place, creating, in the world, the discontinuities that we call states of affairs [...] and to which we refer while thinking and speaking; or on the contrary is it just that these states of affairs, which are already perfectly constituted as phenomena, become the subject of discourse?[1]

In Wittgenstein's *Tractatus* there is a double meaning of the *logical image* (the proposition as 'picture'): on the one hand, it concerns the structural unity of the proposition and on the other, the homology between this structure and the correlated state of affairs. This homology matches the syntactic-semantic connections that constitute the proposition with the real connections that constitute the state of affairs. In this sense, the logical image (*Form der Abildung*), becomes the very form of the appearance of the state of affairs (*Form der Darstellung*). That is why the logical form (*logische Form*) tends to be identified with the form of reality (*Form der Wirklichkeit*).[2] We emphasize the fact, that for Wittgenstein, it is the relation of pictorial similarity between a proposition and the correlated state of affairs which allows us to identify the meaning of the proposition with the evaluation of its truth-conditions.

> The possibility of an object to occur in a state of affairs (its logical form) and that of a proposition to have a truth-value (its form of representation) is part of the possibility that the logical image can be structured parallel to the reality it represents (its form of reproduction).[3]

In other words, contrary to what is the case in model theoretic logic (Tarskian semantics), understanding the relations between language and object depends on the elucidation of the manner in which the structure of a state of affairs can emerge from objective reality. For if it were not so, would the homology between a proposition and the corresponding state of affairs, have any meaning? For Wittgenstein, the structuration of reality into states of affairs corresponds to the manner in which we *think* it. By thinking the reality according to a

1 Ouellet, 1982: 10.
2 See, ibid.: 47–48.
3 Ibid.: 52.

certain state of affairs, we apply to it the corresponding proposition, this projection constituting the form of meaning (*Form der Sinn*). In other words, for Wittgenstein, there is an equivalence between the way in which a state of affairs (conceived as a system of real connections) is manifested and the manner of thinking the meaning of the proposition which supplies its logical image. In this equivalence, we must proceed from manifestation to meaning and not from meaning to manifestation. We must explain these objective 'formal entities' which govern the Thomian 'factors of phenomenological invariance' of the states of affairs. In other words, the thought of the meaning of a proposition must be rooted in the *phenomenological structuration* of reality.

2.5 Structuralism in Phonology (generalities)

In Chapters II and III we will take up the three 'pilars' of linguistic structuralism, namely phonology, structural syntax, and semiotic theory of narratives. But, even at the risk of being repetitive, we will present here their general outlines.

In linguistics, the structuralist perspective goes back to Saussure, especially to the basic concept of *paradigm*. Saussure's main contribution lies in substituting the classical *substantial* criteria of identity with *relational* ones. In a paradigmatic system, the identity of a linguistic unit is referred to as its *value*. It is purely positional. Using a 'geographical' metaphor, we can say that a paradigm is a *categorized* domain D, that is, a domain divided into sub-domains D_i by a system of *boundaries*, K. Each sub-domain D_i is defined by its extension, in other words, by the categorization K. Structure is identified with the *global* organization K, which determines simultaneously the *local* units D_i. Thus, a paradigm is not a system of relations between predefined terms. As regarding their value, the terms of a paradigm do not have any autonomous existence. They can be defined only by their *reciprocal determination*. The category of reciprocal determination is fundamental to structuralism.[1] We recognize here the well known struc-

1 Of course, we should not confuse 'category' in the philosophical sense with 'category' in the sense of a sub-domain of a categorized domain. Similarly, we

turalist 'axiom' as per which difference is prior to identity. Saussure is quite explicit on this point.[1] For him, there are no natural boundaries delimiting the phonetic and the semantic zones corresponding to the signifiers and the signified units of a language.[#] Each term of a paradigm tends to 'occupy' the whole of it, its domain (its value) being limited only by its conflict with the other domains. The definition of a positional value is purely negative, characterized by limiting boundaries. The relations between the terms of a paradigm are relations of dependence in the sense of Sec. 2.3. For Saussure and for semiotics in general, language is a form and not a substance.[2]

Saussure's concept of paradigm was used by Roman Jakobson as the founding concept of phonology. While allophones of a phoneme are substantial units of an auditory-acoustic nature (the units of the substance of expression in the sense of Hjelmslev), phonemes are on the contrary, abstract distinctive units, of a linguistic and functional nature. They can be described as bundles of distinctive features and are governed by phonological rules. Thus, the main theoretical problem is to understand the link between phonetics and phonology, between the organization of the substance of expression and the articulation of the form of expression. It is to understand how phonological categorizations (whose description is the goal of phonology) can emerge as structures from the phonetic substrata, i.e., from the auditory-acoustic flow.

This problem has been seen as a kind of antinomy within general phonetics, and led to a conflict between, on the one hand, *substance-based* reductionist conceptions which regard phonological descriptions as mere artefactual epiphenomena without objective value, and, on the other hand, *form-based* structuralist conceptions emphasizing the ontological autonomy of the form of expression. In the latter perspective, a phoneme is conceived of as

---

should not confuse 'paradigm' in the sense of Saussure with 'paradigm' in the sense of Kuhn.

1 See Ducrot, 1968.

'Signifier/signified' translates the key Saussurian opposition 'signifiant/signifié'.

2 See, Coquet, 1982.

> a differentiating unit having no concrete qualities, but manifested in speech by an allophone having physical (physiological, acoustic, perceptual) qualities which translate into the world of physical realities their differential qualities.[1]

In other words, the form of expression is an *abstract* system which, like the Aristotelian *morphe,* is realized in the substance of expression, i.e., in the concrete event of speech. Now, if we accept that it determines phonetic perception, then we will have to proceed from the abstract to the concrete:

> The description proceeds [...] from the abstract and the functional to the concrete and the material, from form to substance.[2]

But the substance of expression is not an undifferentiated *'hyle'* which would be 'informed' by an ideal form, an essence, an eidos. It is an organized substance. Phonological structuralism should therefore explain how the phonological form can emerge from the organization of the substance. But there is a serious problem here. As Didier Pisoni observed,

> [The] lack of correspondence between attributes of the acoustic signal and the units of linguistic analysis has been and still currently is, one of the most important and controversial issues in speech perception.[3]

The key for the resolution of this difficulty is to be found in the structure of phonological perception whose essential property is to be what is called *categorical*. This means the following.[4] Studies on the structure of speech sounds have shown that it depends on a small number of parameters, called *acoustic cues,* that can be varied continuously in speech synthesis. Tests of identification and discrimination reveal that discrimination is subordinate to identification. In other words:

(i) identification categorizes (discretizes) the continuous space of acoustic cues, and divides it into domains corresponding to stable perceptions, and

---

1 Malmberg, 1974: 220.
2 Ibid.: 30.
3 Pisoni, 1979: 334.
4 See, Petitot, 1982b, 1983b.

(ii) there is no intracategorial discrimination.

It is this second feature that defines phonetic perception as categorical and distinguishes it from continuous perception where the discriminating capacity is essentially independent of categorization. It allows us to understand how perception can spontaneously *discretize* the auditory-acoustic flow, or in other words, how discontinuity can emerge from continuity. In this sense, it establishes a link between the audio-acoustic level of phonetics (organization of the substance of expression) and the linguistic level of phonology (abstract relational nature of the form of expression): the phonemes encoded in the auditory-acoustic flow are categorical as a consequence of the perceptual process itself; they have a psychological reality as discrete units.

2.6 *Actantial structures and case-grammars (generalities)* [1]

In syntax, the structuralist approach goes back to Lucien Tesnière.[2] For Tesnière, a sentence is essentially a system of *connections* which, being 'disembodied', exist only in the 'mind'. The structural connections are not of a symbolic essence, but are part of a 'vital and organic' principle of organization (Tesnière refers to Humboldt's celebrated *innere Sprachform*). Tesnière schematized them by means of graphs called *'stemmas'*, which can be considered as the precursors of the syntactic trees used in most formal linguistic descriptions since Chomsky. As the visual manifestation of abstract dependence relations (see Sec. 2.3) a stemma is nothing but the structural schema of a sentence.

For Tesnière, the structural connections define the *functions*, that is, the *roles* assigned to words in the expression of thought. They are projected on the linear order of syntactic concatenations, and structural syntax is therefore dependent on 'the relations between the structural and the linear orders'.

Recent developments in transformational-generative grammar and generative semantics might suggest that Tesnière's original structuralist position has been satisfactorily formalized and is now

1 For more details, see, Petitot, 1982c, and Chapter II.
2 See, Tesnière, 1959.

obsolete. But, that is not the case. Actually, these trends represent a static, taxonomic, formalist, and logico-combinatorial conception of syntactic structures, rather algebraic, and very different from Tesnière's dynamical, 'vitalist' and Gestaltist viewpoint. Indeed, Tesnière has always emphasized that syntactic structures are self-regulated organizations akin to biological organisms, that structural syntax is neither a logically nor a psychologically based grammar, and that it is functional and dynamic and not categorial (in the sense of the grammatical categories) and static. In fact, his conception is an *actantial* 'scenic' one based on the notion of *verbal valence.*

It was only with the *case grammars* of Fillmore, Chafe and Anderson, and later the *relational grammars* of Keenan, Comrie and Johnson based on the works of Perlmutter and Postal, and still more recently the *cognitive grammars* of Talmy, Langacker and Lakoff, that this conception of syntax received renewed attention. In 'The Case for case reopened',[1] Charles Fillmore reintroduced a scenic conception of syntactic structures for the following reasons. Classical case grammars, in spite of their early success, ran into serious difficulties, particularly on the question of defining the case universals, because they are based on a *semantic interpretation* of deep syntactic relations. Their basic hypothesis is that there exist a finite set of deep case-universals which are also functional categories (e.g., Agent, Dative, Instrumental, Locative, Objective and Beneficiary), whose *notional* content can be determined, which select the semantic (actantial) roles, and which, though of a semantic nature, can be discovered applying purely *syntactic* criteria. But from a cross-linguistic comparison, one can conclude that there exists a conflict between the proliferation of cases entailed by their conception as classifiers of sentences, and their limitation entailed by their conception as universals. If case universals are assigned a distinctive notional content, then they will have to share the lexical content of the verbs, and thus they will proliferate. Instead, if they are given a sufficiently broad notional content so as to form a restricted set (a closed class in Talmy's sense), then their content will become too broad.

1 Fillmore, 1977.

A first solution to this difficulty was proposed by John Anderson [1] on the model of the distinctive features analysis of phonemes. It consists in:

(i) treating case-meanings as complex contents analyzable into case features ('multi-case' analysis of the actantial roles);

(ii) classifying case features into a limited number of universals, on the basis of the *localist hypothesis*, according to which the *positional relations between spatio-temporal actants operate as schemas for the actantial relations;*

(iii) positing that verbs select case-features;

(iv) elaborating a 'generative grammar' of such selections.

Fillmore's solution is different. It is based on the observation that several different semantic fields can form the substrata for a single abstract schema of actantial connections. Fillmore introduces within case semantics a distinction between the specific semantic field under consideration and the purely positional meanings defined by the actantial stemma. He calls these semantic fields, 'scenes'. Each 'scene' is lexico-syntactically organized by a restricted number of specific constructions which select the corresponding cases with appropriate syntactic and lexical features. Hence the slogan: 'Meanings are relativized to scenes'.[2]

In developing this scenic conception, Fillmore employs the notion of *case-frame* which serves as an intermediary between the description of situations and the underlying syntactic representations. A case-frame assigns semantico-syntactic roles to the actants of the process represented by the sentence and thus constrains the choice of a perspective that selects one of the actants as the grammatical subject in relation to a case-hierarchy. Fillmore is arguing for a *conceptual* definition of cases in placing them at the interface of language and thought: we produce and understand linguistic expressions by 'activating' in our mind prototypical scenes. In other words, when perspectivized, an expression evokes the global background on which it is profiled.

---

1 See Anderson, 1971; 1975a, b.
2 Fillmore, 1977: 59.

Thus in structural syntax, we again encounter all the problems of Gestalt theory and phenomenology that we have already referred to (Sec. 2.3 and 2.4): linguistic universals should be rooted in the perceptual organization of the state of affairs.

The most delicate issue is however to arrive at what we will call *a configurational definition of case meanings.* Indeed, even if case meanings are relativized to scenes, they still continue to have a purely positional value. Now, their notional (actantial) content cannot be defined as autonomous, but only in terms of *relative positions* in actantial schemas. These positions are reciprocally determined as paradigmatic values. Structural syntax thus has to tackle a double problematic:

(i) how can actantial schemas emerge as self-regulated structures, dynamic morphologies, and syntactic Gestalten, from the phenomenological organization of reality into states of affairs?

(ii) how do these schemas allow for the description of case meanings in terms of positional values?

To solve these problems, it seems necessary to resort to the *localist hypothesis*,[1] and assume that structural connections between spatio-temporal actants can serve as model for syntactic schemas in general. Such a hypothesis has been thoroughly confirmed by René Thom's interpretation of case universals in terms of elementary catastrophes. Most of the theoretical difficulties that we have indicated above can be solved from this perspective.[2]

2.7 *Semio-narrative structures (generalities)* [3]

Actantial structures exist not just at the sentence level. They are also present at the level of *narratives,* as demonstrated in the structural analysis of folktales. At the surface (manifest) level, myths, fairy-tales, drama, novels, etc., relate intrigues involving *dramatis personae* (the actors) who are situated spatio-temporally, defined by thematic roles, linked by kin relations, and interacting through conflicts, gifts,

1 For an historical account of the localist hypothesis, see Chapter II and Hjelmslev, 1935.

2 See, Petitot, 1979c, 1982c; Wildgen, 1981, 1982; see also Chapter II.

3 For more details, see Chapter III.

contracts, separations, unions, passions, etc. In spite of the proliferation of the superficial discursive structures, we can identify certain deep structures, which A.J. Greimas called *semio-narrative*. This method of analysis is able to show that:

(i) the abstract (non-figurative) discourses, be they philosophical, political, or scientific are also partly organized on similar bases;

(ii) the deep semio-narrative structures reflect also *lived* experiences of passions, ideologies, actions, dreams, etc.

This suggests that *semio-narrative structures are anthropological structures of mind*.

Greimas' semio-narrative grammar is mainly concerned by a truly original relationship between syntax and semantics, namely the projection (or conversion) of the paradigmatic axis onto the syntagmatic axis which constitutes one of the central thesis of structuralism.

The recognition of a syntactic component of deep semio-narrative grammar [1] goes back to Propp's analysis of Russian folktales in his celebrated work *Morphology of the Folktale*. Underlying the actions of the characters in a tale, Propp identified a set of *functions* (that is to say, typical actantial relations) *canonically* ordered, appearing in a rule-governed manner as if in a process of morphogenesis, and expressed by typical sequences: establishment of an initial lack (e.g. by transgression of social rules, deception, etc.); contract between a Sender (e.g. a king, a dominant social group, etc.) and a Hero; a series of tests, first a qualifying test by which the Hero acquires the modalities of wanting, knowing and/or being able to (e.g. obtaining a magical instrument from a Helper), followed by a decisive test (main test) wherein the Hero accomplishes a feat that liquidates the initial lack (e.g. killing a dragon), and finally a glorifying test in which the Hero's performance is approved by the Sender. In other words, Propp identified invariant, stable, and universal actantial structures governed by an actantial syntax which syntagmatizes an actantial *paradigm* consisting of typical actants such as Sender/Receiver, Subject/Object-of-value, Subject/Anti-subject, Helper/Opponent.

---

1 Semio-narrative grammar has at least two levels: deep and surface levels. Each level has two components: the syntactic and semantic ones.

After Propp's syntactic achievement, it was Lévi-Strauss who introduced a *semantic* component in deep semio-narrative grammar.[1] It is indeed the most challenging part of structural narratology. This deep semantic component is very different from the discursive and figurative surface grammar which distributes lexical meanings along the surface syntactic structures. Of course, one can analyze narratives like myths by focusing only on the discursive-figurative 'clothing' of deep actantial syntax, but then one will not understand their anthropological function, except, as is often done, by interpreting the surface contents symbolically. It is precisely this anthropological function that Lévi-Strauss sought to define. For demonstrating the logical coherence of myths, he was committed to a semantic approach. This was the basis of his criticism of Propp. Lacking a correct understanding of the relationship between the paradigmatic and the syntagmatic dimensions, Propp was only able to identify the syntagmatic invariants of an actantial syntax.[2]

Inspired by the Prague school (Jakobsonian) phonology and Hjelmslev's principle of the parallelism between the expression plane and the content plane, Lévi-Strauss introduced in the theory of deep semio-narrative structures, the most important aspect of the paradigmatic dimension, namely *categorization* (see Sec. 2.5). The idea is that tales, and particularly myths, share a level of deep semantic categorizations expressing values which belong to unconscious codes (familial, natural, cosmological, economic, culinary, etc.) and which are projected on the syntagmatic dimension. This deep semantics does not correspond to the surface lexical meanings. It has a contextual and global function. It selects from the surface lexical figures ('sememes') certain specific semantic features ('semes'). But the contents it articulates (Life/Death, Nature/Culture, Man/Woman, Divine/Human, etc.) do not have a reference in the objective world. They are some sort of psychical drives or ideals that 'give meaning to life', *a meaning that cannot be grasped as such but only experienced via its conversion into actantial structures*. More precisely,

1 See, Lévi-Strauss, 1958 and 1964–1971.

2 See, Coquet, 1982.

(i) the deep semantic categories are anthropological universals of the imaginary order;#

(i) they act only when axiologized and ideologically invested in the objects of value, the quest for which motivates the actions (the 'narrative programmes') of the subjects;

(iii) only the circulation of such objects of value can allow them to be subjectivized; in other words, they can become part of the subjects only through experiences and actions;

(iv) thus, actantial syntax *converts* the deep semantics of the tale into a narrative 'doing' which determines its anthropological function. It allows to grasp the unconscious structures of subjectivity, by simulating a 'theatre' that presents them in a scene.

From this point of view, we can see that a correct semio-narrative grammar would involve the resolution of three problems:

(i) what is an elementary narrative structure (we already know that it is of an actantial nature)?

(ii) what is meant by semantic categorization (we already know that it is similar to categorization in phonology)?

(iii) what is the nature of the conversion of deep semantics into syntax (we already know that it is a projection of the paradigmatic axis onto the syntagmatic one)?

These are, among others, three questions that Greimasian theory is mainly concerned with.

As regards deep semantics, we must first of all *formally* define the morphologies which constitute the categorial *form* of content (in the sense of Hjelmslev). This is the function of the *semiotic square*.[1] According to Greimas, the semiotic square is a universal schema for the articulation of meaning, for the apprehension of which it ensures the minimal conditions. As an elementary morphology prior to any sememic investment, it unfolds a *semic category* connecting two

---

In French, it is very easy to substantivize adjectives or verbs for constructing abstract nouns: *l'imaginaire, le symbolique, l'être, le faire*, etc. As it is not the same in English, we will use expressions such as 'imaginary order', 'imaginary stance', or 'imaginary function'.

1 For some general reflections on the semiotic square, see, SES, 1976, and BGRS, 1981.

contrary semes into a relation of *junction* (conjunction/disjunction as reciprocal presupposition).

Taken simply as a logical form in the framework of elementary Boolean logic, the semiotic square is completely trivial. It is only the reformulation of the 'logical squares' whose tradition goes back to Aristotle. But everything changes if we observe that it is a structure in the strong sense, that is, an 'organic' and 'self-regulated' system of dependence relations (see Sec. 2.3) defining positional values. The relations of contrariness and contradiction which are constitutive of it are not logical. As shown by Arild Utaker,[1] they are respectively equivalent to *qualitative* and *privative* oppositions in Jakobson's sense. They have therefore to be treated as such.

Now, the notions of opposition and reciprocal presupposition rely on a primitive notion of position which is primarily *topological* and not logical. Similarly, oppositions are based on conjunctions and disjunctions and these notions rely on a primitive notion of junction which is also primarily topological, and not logical. Thus the formal essence of the semiotic square, depends ultimately on a dynamical topology of places and connections and not on a static logic of terms and relations. We have shown that such a framework is provided by the elementary catastrophes.[2]

As for the narrative syntax, Greimasian theory regards elementary actantial structures as the syntagmatization of the paradigmatic relations constituting the actantial model (narrative programmes). These relations are basically reduced to relations of reciprocal presupposition between Sender and Receiver (contract), Subject and Object (acquisition of modal competence and/or performance), Subject and Anti-subject (performance). As regards the subject/object relation, Greimas views the subject as an *intentional* subject (a subject of lack, of quest, of desire) persuing a semantic value invested in an object. This means that the basic narrative programme is to realize a conjunction between a subject and an object of value.

There are, however, some difficulties. For instance, Paul Ricoeur has pointed out that the phenomenology of action unfolded in narratives cannot be reduced to a mere syntactic 'doing' consisting of

1 Utaker, 1974. See also, Section III.3.3.

2 See, Petitot, 1977b; 1982a.

simple operations of conjunction and disjunction between subjects and objects of value.[1] Further, the conception of semiotic subjects as intentional subjects, evidently raises the question of the nature of their intentionality. A closely related issue concerns the problematics of belief, seduction, manipulation, and selection of objects of value.

But the main difficulty is still the conversion of deep semantics into actantial syntax, i.e., the projection of deep semantic categories onto narrative programmes.[2] In Greimas' theory, it is tackled by the introduction of an intermediate level, the 'syntax of operations', *akin to both the semiotic square and the actantial structures*. The guiding principle is that the constitutive relations of position and junction can be converted into logical operations on the terms they define, and that these operations can themselves be accounted for by actantial interactions, i.e., by a 'syntax' of action.

Though partially acceptable, this response to the problem of conversion remains incomplete and has to be further interrogated. It clarifies neither the nature of intentionality, nor the mechanisms by which an unconscious drive can be invested in an object and confer on it the 'aura' of an object of value. But these phenomena go far beyond pure structural semiotics. Their comprehension would require a 'metapsychology', either in the Freudian sense, or in the sense of an anthropology of the imaginary stance.[3]

3 The problem of formalizing structures

3.1 *The intrinsic limits of the formalist perspective*

This overview of some key aspects of structuralism shows that, as a conceptual and methodological perspective, it is intrinsically transdisciplinary (touching upon biology, anthropology, Gestalt theory, cognitive psychology, phonology, linguistics, semiology) and that in

---

1 See, Ricoeur, 1980; and Chapter III. 4.

2 For a description of certain aspects of this conversion, see, BGRS, 1982.

3 See, for instance, Brandt, 1982a; Thom, 1981, 1983; Petitot, 1982g.

each of its domain of empirical validity, it reveals, as emphasized by Piaget, 'a common positive ideal of intelligibility'.[1]

As an epigenetic and relational doctrine of organization, structuralism represents, along with physics, practically the only area where several different domains are brought to a rational synthesis. The question of its *formalization* is therefore especially crucial.

Now, it can be observed that structuralism, owing to the lack of suitable formal tools, never reached its intended objectives. Until now, it only succeeded in developing a symbolic (logico-combinatorial) conception, a conception we are going to criticize, and to which we shall propose an alternative.

The symbolic perspective in structuralism is twofold. It is concerned with formalization of either structuralist conceptual theories or structures conceived as empirical phenomena.

As regards the formalization of structures viewed as a particular class of phenomena, the situation may appear quite satisfactory. Indeed, the structural methodology has resulted in a plethora of *models*. In fields such as system theory, cybernetics, artificial intelligence, language automata theory, formal grammar, categorial grammar, intensional logic, etc., it has been quite successful. We must, however, note that these achievements were made possible by a general 'reification' of structures, enabling them to be 'algebraized'. Dynamical structuralism, which, for us, is the genuine structuralism, has meanwhile remained in the dark. In biology, for instance, whatever be the usefullness of formal cybernetic schemas, their application encounters, as Thom notes,

> very serious difficulties as soon as we move from an abstract schematism to a material realization in space-time.[2]

That is, from a formalist perspective, one cannot always understand how the structures represented by such schemas can emerge from the physico-chemical properties of the substrata. This perspective is valid

> only for partial mechanisms, ready-made, and in full functional activity. [...] In no case it can be applied to the global structure of living beings, to their epigenesis, or to their physiological maturation.[1]

---

1 Piaget, 1968.
2 Thom, 1980a: 154.

That is why, in biology, it is pertinent to partially resort to a morphodynamical approach which allows us to understand the emergence of structures out from their physico-chemical substrates.

In linguistics and in semiotics, formalism has similar intrinsic limitations. The widely accepted belief 'that the only generative virtue of a structure, coming from its pure form, should be admitted *a priori*, and needs no explanation'[2] should be questioned, for, in the case of natural language,

> it is the self-limitation of generative capacities of syntax that requires explanation.[3]

In order to understand this self-limitation, we have to consider the *dynamics* underlying the formal cinematics described by formal languages. We have to focus on structures based

> [not] on the automatic iteration of certain operations, but on the contrary, on an intrinsic combinatorics provided by the dynamical interpretation.[4]

As regards now the formalization of structuralist theories, there is a lot of confusion, due to the strong influence exerted by mathematical structuralism. The idea goes back to Husserl who attempted to 'axiomatize' in a quasi-Hilbertian style a formal ontology of dependence relations. Later, it was adopted by Hjelmslev in semio-linguistics and became the basis of Greimasian epistemology.

From this perspective, a theory is considered as a conceptual system, a descriptive metalanguage, defining hierarchically concepts from primitives. Being undefinable, the primitive concepts behave like *regional categories*. In Greimasian theory, the primitives are notions such as continuous/discontinuous/discrete, relation, difference, opposition, junction (conjunction/disjunction), transformation, operation, etc. It is assumed that by providing them with a formal expression, it will become possible to 'axiomatize' the descriptive

---

1 Thom, 1972a: 207.
2 Thom, 1980a: 164.
3 Ibid.
4 Ibid.: 106.

metalanguage and convert it into a formal language, a 'pure algebra.'[1]

The main difficulty with this perspective is that it can, at best, only elucidate the logical architecture of the theory and cannot obtain an effective *mathematization of its content.* It relies upon a formal logic concerning the linguistic form of knowledge and not a transcendental logic concerning the *objective* content of knowledge. It misunderstands the critical fact that, in science, mathematical schematization of regional categories is the key to any constitution of objectivity and consists not in an 'axiomatization', but in a mathematical *interpretation* of the categorial content. In this way, *there exists a conflict between formalization and mathematization in the structural field.* This point is for us of utmost importance. We will see that it is the topological and dynamical interpretation of the structuralist categories which constitute the basis of their objective significance.

3.2 *The topological a priori as the central theoretical problem of structuralism*

In all the domains we have considered, structural categoriality is always the same and possesses a content which ultimately refers to topological intuitions (position, junction, paradigmatic categorization, connection, etc.). As we have seen, every structure is primarily a structurally stable and (self-)regulated system of connections between positional values. This basic fact should guide any attempt of theorization in this field. Let us therefore further clarify the nature and the significance of some of the issues we have already mentioned.

One of the main hypotheses of structuralism is that the paradigmatic dimension of systems constrains their syntagmatic one. Now, logical formalisms (elementary formal logic, modal logic, intensional logic, combinatorial logic, algebras of relations, automata, category theory, topoï, etc.) which are used generally to formalize the semantic and/or syntactic descriptions, involve an elimination of the paradigmatic axis in favour of the syntagmatic one. They are therefore unsuitable for structural formalization. Whence a first formulation of the central theoretical problem: *what kind of formal essence*

1 Greimas-Courtès, 1979: 225.

should be attributed to the paradigmatic dimension of semiotic systems in general?

The structuralist axiom posits that the paradigmatic organization is purely relational and determines abstract units which possess no independent identity, and exist only as pure positional values. It asserts the primacy of difference over identity in the semiotic realm. Whence a second formulation of the central theoretical problem: *what kind of mathematical content should be assigned to the category of relation in such a manner that it could schematize the structural primitives of difference and positional value?*

There is actually an irreducible gap between structuralist eidetics and symbolic formalisms. In its pursuit for a symbolic calculus and a control of the recursive complexification obtained by iterating rules, the latter ignores the evident fact that, in the case of natural language, the relations are *meaningful* relations belonging to the *form of content*. As Greimas emphasized, this is one of the 'fundamental options' for structuralism:

> while the symbol-units of a formal syntax constitute an alphabet (i.e., some sort of an inventory, often wrongly referred to as 'structure') governed by a set of operational rules, the units of a conceptual syntax are arranged in a taxonomy (a sort of elementary morphology) upon which the syntactic operations are performed.[1]

The distinction between formalist theories of syntax, which develop a symbolic calculus of recursive linguistic properties, and conceptual theories of syntax, which model the paradigmatic relations of the form of content now appears as a true antinomy.

The paradigmatic organization of semiotic systems involves a taxonomic dimension. But in its structuralist reworking the classical concept of taxonomy undergoes a mutation. Traditionally,[2] taxonomy is concerned with the classification of already defined, individuated, and autonomous objects. In structuralism, on the contrary, the abstract units are defined and determined by the classification itself. Taxonomy concerns therefore the emergence of discrete units from a continuous substratum by a process of categorization. That is why the primitive notion of discontinuity – of difference –, is in some

1 Greimas-Courtès, 1979: 378. We will come back to this quotation in Sec. II.2.4.
2 See, for instance, Gil, 1981.

ways the 'pure intuition' of the structural order. A structural 'space', we have seen, is a continuous space categorized by a system of discontinuities and thus discretized.

The paradigmatic dimension is then the new apellation for the taxonomic one when we no longer assume that a multiplicity of discrete units already individualized are distributed in an abstract system of equivalence classes, but on the contrary, that a categorizing classification discretizes a substance and defines discrete units by reciprocal determination.[1] In a paradigm, the positional values of the units result from a process of *morphogenesis.*

In this new perspective on paradigms, the central theoretical problem *is to work out an adequate concept of space.* A 'structural' space would evidently not be a physical one. It is a generalized space of deformable entities (for instance, semantic units or acoustic images), a functional space of internal states of a 'black-box'.

The hypothesis of a spatiality immanent to the paradigmatic dimension explains the mutation undergone by the taxonomic one within structuralist practices. It involves a conception where the taxonomic structure becomes *a synthesis between the concept of classification and the concept of generalized space.* Paraphrasing Riemann's well-known statement on the concept of manifold [2] we can say that: within a discrete manifold the principle of the relations between units is already present in the concept of this manifold while, in a continuous manifold, this principle should come from outside. It is therefore the case that either the reality on which the classification is founded forms a discrete manifold, or the basis of the relations are to be sought outside of it, i.e., in the form which categorizes it.

The theoretical challenge is then to *geometrize* paradigmatic categorizations. We can reformulate the central theoretical problem as follows: how can a geometrization of the paradigmatic synthesis between the concepts of classification and generalized space be used

1 See Gil, Petitot, 1981.

2 Riemann, 1854. 'In a discrete manifold, the principle of metric relations is already contained in the concept of this variety, while in a continuous manifold, this principle should come from outside. It must then be the case that either the reality on which the space is founded forms a discrete manifold, or the foundation of metric relations be sought outside of it, in the binding forces which work within it.' (Our translation).

(i) to assign a mathematical content to the structural primitives of difference and positional value;

(ii) explain the differentiating action of discontinuities, which results in the formation of discrete units, and

(iii) deduce, at the syntagmatic level, the conceptual syntactic relations manifesting the form of content?

It is the resolution of these difficult questions – which go far beyond a simple descriptive-conceptual theory – that is the original task of a mathematical schematization of structures. Such an imperative is so alien to the social sciences that it is remarkable it could have been philosophically formulated very early: we have in mind a major work of Gilles Deleuze.

3.3 *Deleuze's proposal for a schematism of structure*

Deleuze's reflections on structuralism should be understood from a philosophical standpoint. His project is to evaluate structuralism in showing how it modifies the transcendental tradition. In his article 'A quoi reconnaît-on le structuralisme?'[1] that we summarize below, he analyzes the different structuralist approaches according to seven criteria.

3.3.1 *The symbolic realm*

> The first criterion of structuralism is the discovery and the recognition [along with the imaginary and the real orders, but deeper than both] of a third order, a third realm, that of the symbolic one. (p. 301) [2]

The symbolic stance (recognized for the first time by structural linguistics) is the 'element' of structures. It is very difficult to be decerned as such, because it is always hidden by the concrete properties of the substrata where the structures are implemented in. A

1 Deleuze, 1973. The page numbers will be referred to in the text.

2 The triad 'real, imaginary, symbolic' was one of the main themes in the sixties. In this context, the term 'symbolic' has nothing to do neither with 'symbolic' in the logical sense, nor with 'symbolic' in the mythological sense.

structure is neither a Gestalt, nor a figure of the imagination, nor an intelligible essence.

> It is a combinatorics of formal elements which possess in themselves neither form, nor signification, nor representation, nor content, nor empirical reality, nor hypothetical functional model, nor intelligibility behind the appearances. (p. 303)

3.3.2 *The criterion of locality or of position*

If the symbolic stance refers to no pre-existing reality, if no imaginary or conceptual content provides it with a signification, if 'the elements of a structure have neither extrinsic designation nor intrinsic signification', it is because

> as Lévi-Strauss had clearly pointed out, [these elements] have nothing but a *content: a content that is uniquely and necessarily 'positional'*. (p. 304).

This is indeed the most crucial fact. We shall do well to meditate on Deleuze's assertion that

> the scientific ambition of structuralism is not quantitative but topological and relational (p. 305).

We must emphasize here the term 'topological'. As Deleuze insists:

> what is structural, is space, but a non-extended and pre-extensive space, pure spatium (p. 305).

To give a scientific status to the 'and' in the expression 'topological and relational' constitutes the main challenge. That is why

> structuralism is inseparable from a new transcendental philosophy wherein the places are more important than the things that fill them (p. 306).

The consequences of the locality criterion are far reaching. The content resulting from the combination of purely positional identities is always an effect 'produced as an excess by the combination of places in the structure', it is always overdetermined (p. 306). That is why the symbolic order transcends both the real and the imaginary ones:

> Places in a purely structural space are prior to the things and the real beings which occupy them, prior also to the imaginary roles and events which necessarily appear when the places are occupied. (p. 305)

3.3.3 *The differential and the singular dimensions*

The 'pure intuition' of structuralism can thus be easily formulated. It is the intuition of dividing a 'substratum space' by means of a system of thresholds. The problem is to make such systems autonomous relative to the real and/or semantic identities which they are invested with. It is to conceive articulations of differences that are *independent* of substrata though existing only if implemented in them. In that sense, structuralism relies upon a very typical *functionalism*. It can even be considered as the true source of functionalism.

Such a conception of the symbolic stance requires a 'geometrization' of the primitive notions of position and difference. This is required firstly, for freeing the positional values from any purely logical principle of identity (the symbolic effects are due to an opposition between positional identities and real or semantic identities), and secondly, for viewing differences as resulting from a *genetic* process of differentiation.

Structural objectivity thus hinges on a 'geometry of position' whose mathematical essence must be understood. Deleuze remarkably anticipated its general idea while taking recourse to the mathematical notion of *singularities* of differential equations:

> The important notion of singularitiy seems to be relevant in all domains where one speaks of structure.

A structural 'space' is a space divided into regions (places) by a system of differences. It is a space of coexistence, of colocalisation. For the symbolic elements that it is invested with, the differences are not external relations between pre-existing identities but constitutive relations of 'reciprocal determination'. Genetically, they are obtained from the *unfolding of singularities*. When singularities unfold in 'space' they differentiate and organize it into a 'structural space'.

> The *reciprocal determination* of symbolic elements extends to the *complete determination* of the singular points which constitute a space corresponding to these elements (p. 309).

> Every structure reveals two aspects: a system of differential relations on the basis of which the symbolic elements are reciprocally determined, and a system of singularities corresponding to these relations and tracing the space of the structure (pp. 309–10).

A domain can be defined in structural terms if:

(i) symbolic elements are 'embodied' in its objects;

(ii) differential – i.e., categorical – relations are 'actualized' in the real relations linking these objects;

(iii) singularities – i.e., 'events' – 'distribute' roles and functions to the objects which occupy them (p. 310).

In this sense 'every structure [...] represents a category-function complex' (p. 311).

3.3.4 *Differentiating and Differentiation*

The main difficulty encountered by the structuralist *épistémé* is that structures are 'masked by their products or effects' (p. 316). The expression of a structure is always a fading of its symbolic nature. For a structure is never actualized as such. Being 'real without being actual, ideal without being abstract', pure 'virtuality of coexistence pre-existing to the entities' (p. 313), a structure is 'embodied.' It is actualized in the production of spatio-temporal and/or semantic differences but, on being actualized, it vanishes as such (see, Sec. 1).

To underline this specific manner in which the metaphysical opposition virtual/actual acts on the concept of structure, Deleuze plays with the terms *differentiation/differenciation* (p. 314). In so far as it is virtual, a structure is differentiatal. On actualizing itself it becomes a principle of differenciation. A structure 'is differential in itself, and differenciating in its effect' (p. 315).

3.3.5 *The serial function*

After having identified the non-trivial theoretical content of structuralism, Deleuze goes on to show that structuralism is also non-trivial in practice. Since Lévi-Strauss' account of totemism, we know that a symbolic system of differences (e.g., a zoological taxonomy) can be used for encoding another symbolic system (e.g., social relations). The practical content of structuralism lies in showing how, by projecting the paradigmatic axis on the syntagmatic one, the symbolic elements of a structure 'are serially ordered' and how a series always refers homologically to another series (p. 318). According to Deleuze, Lévi-Strauss' main contribution is in showing that this type of homology between two series is not a trivial encoding, that is, a mere term-to-term correspondence. In fact, the places (positional identities) in the first series are inseparable from *displacements* induced by the second series (p. 320). The problem is of explaining how the relative displacements in the series 'are absolutely parts of the places in the structure' (p. 321). It is concerned of solving the paradox of a *metonymic* principle of identity for positions. The identity of a symbolic place is not what ensures its stability, but what ensures the possibility of its displacement. Metonymy violates common sense logic.

> This relative displacement of the two series is not at all secondary; it does not affect their terms from outside and secondarily as if to give them an imaginary disguise. On the contrary, displacement is strictly structural or symbolic: it belongs essentially to the places in the space of the structure, and thus governs all the imaginary disguises of the beings and objects which happen to secondarily occupy these places (p. 321).

3.3.6 *The empty place*

If the relative displacement (metonymy) can be an intrinsic part of the identity of position, it is because every structure 'contains an object or an element which is quite paradoxical' (p. 321). This paradoxical element is of a kind very different from the symbolic elements, the differential relations, and the singularities. It circulates within the series as if it was 'its own metaphor and its own metonymy' (p. 322). It lacks any ontological function (it is not an object), any self-likeness (it is not an image), any logical identity (it is not a concept) (p. 323). And if the relative displacements is an intrinsic part of positional identities, it is because the relative place of the terms in the structure depends on their absolute place in relation to this element.

> It is in this sense that displacement, and more generally any form of exchange, is not something coming from the outside, but the fundamental property which allows one to define structure as an order of places under the variation of relations (p. 324).

3.3.7 *From the subject to practice*

When a structure is actualized, real and/or semantic entities occupy its places. But the places are already virtually occupied by symbolic elements which determine their colocalisation. Nevertheless the 'empty place' escapes this 'primary symbolic filling-in' (p. 330). Being of a nature different from the symbolic elements, the differential relations and the singularities, the 'empty place' remains empty. 'Being its own symbol, it does not have to be filled-in' (p. 330).

And precisely because it remains empty, it is the 'metonymic' principle of identity of the symbolic elements, and is correlated with the 'eminently symbolic' instance which the *subject* is (p. 330). This subject is 'symbolically affected' by the 'ideal events which form part of the structure itself', that is, by *immanent* events in the structure (p. 332). It is in this sense that structuralism is also a praxis (p. 333).

Deleuze's criteria of the structuralist *épistémé* tends towards a positional schematism of the category of relation. It asserts the necessity of deducing the 'logic' of meaning from a primarily spatial conception. It helps us to understand that a 'logic' of meaning must be in fact, a 'physics' of meaning. *The symbolic order is to the semantic substance what morphogenesis is to matter.*

4 The necessity for a morphological geometry

The structuralist problematic has a philosophical genealogy. If, as Jacques Derrida once claimed:

> *stricto sensu*, the notion of structure refers only to space, a morphological or geometrical space, i.e., an order of forms and places [1]

1 Derrida, 1967: 28.

and if Deleuze's characterization of structuralism is valid, then the schematization of structural categories depends entirely on the possibility of mathematically determining the 'positional geometry' which operates as the 'form of intuition' for structural phenomena. It depends *hic et nunc* on the elaboration of a true *geometry of position*.

Now, as Buffon emphasized long ago, regarding embryogenesis, such a geometry of position has always been radically missing:

> Whatever is directly connected with position is totally lacking in our mathematics. This art which Leibniz called *Analysis situs* is yet to be born, and still, this art which would let us know the relations of position between things would be as useful and perhaps more necessary to the natural sciences than the art which only account for the quantitative aspects of things; for, often it is more important to know about form than about matter.[1]

Leibniz himself wrote on February 2, 1706, in a letter to Rev. Fr. des Bosses:

> If we assume the fullness of things (as the Cartesians do) and the uniformity of matter, and if we introduce just movement, then we always get a sequence of equivalent things; [...] and thus, nobody can distinguish the state at one moment from that at another, not even an angel; in this way, one would never find any variety in phenomena: hence in addition to figure, magnitude, and movement, one must admit forms by means of which the difference of appearances emerge within matter, forms that one can grasp intelligibly, it seems to me, only from Entelechies.[2]

Concerning the possibility of constituting a *geometrical descriptive eidetics*, that is, a geometry of morphological types adequately described by the morphological concepts of natural language, Husserl in his turn declared in *Ideen I* (Sec. 71–75):

> The geometer is not interested in the forms given in sensible intuition, as does a scientist in a descriptive study of nature. He does not construct, like the latter, *morphological concepts* bearing upon vague types of forms which, being founded on sensible intuition, could be directly grasped and whose concepts and terminology would be as vague as these types themselves.[3]

---

1 Buffon, 1744, t. IV, Chap. IX, p. 73.
2 Leibniz, 1706.
3 Husserl, 1913: §74 "Contrast between geometry and descriptive science".

> The most perfect geometry and its most perfect practical mastery can hardly be of help to the philosopher who wants to describe nature for expressing with exact geometrical concepts the things he actually expresses in an extremely simple, comprehensible, and fully appropriate manner, by using words like serrated, notched, lens-shaped, umbellate, etc.; these simple concepts are *inexact in essence, and not by chance; for the same reason, they are also non-mathematical.*[1]

> Whatever be the achievements of an exact science, that is a science operating with ideal understructures, it cannot solve the originary and entitled tasks of a pure description.[2]

Thus structuralism, as far as its mathematization is concerned, requires the elaboration of a general mathematical theory of morphologies and morphogenesis. Catastrophe theory is the first effective proposal in this direction, and that is why it is crucial in this endeavour.

5 The principles of Catastrophe theory

At the beginning of his pioneering work *Biology and Structuralism*, René Thom asked:

> Can the recent structuralist trends in social sciences such as linguistics and anthropology provide new methods for the edification of an experimental science like Biology?[3]

In other words, is it possible to reach a synthesis between the dynamical structuralism focusing on morphogenesis and the phonological or semiolinguistic structuralism focusing on the form of semiotic systems? I think we can now answer this question positively. For this, we have:

---

1 Ibid.
2 Ibid.
3 Thom, 1968b.

(i) to reduce every structure (paradigmatic categorization, actantial interaction, morphogenetic differentiation, etc.) to a morphology defined on a suitable substratum space;

(ii) to reduce every morphology to a system of qualitative discontinuities on its substratum space.

We could then seek to describe the observed morphologies in the chosen empirical corpus as gluings, combinations, or concatenations of a small number of structurally stable and recurrent sub-morphologies, what are called 'morphogenetic fields' or 'chreodes'. We could further seek to elaborate firstly *local* dynamical models that generate these chreodes, and secondly *global* dynamical models capable of explaining purely morphologically the stable associations of chreodes as well as the phenomena of order and hierarchy.

Before concluding this introductory chapter, we will then briefly outline the basic principles of catastrophe theory (CT).

5.1 *Phenomenology and objectivity*

Early in *Structural Stability and Morphogenesis*, René Thom assigns to CT the task of explaining the stability, the transformation and the succession of forms.

> One of the central problems posed to the human mind is the problem of the succession of forms. Whatever be the ultimate nature of reality (assuming that this expression has any meaning), it is undeniable that our universe is not a chaos; we discern in it beings, objects and things that we denote by means of words. These beings or things are forms, or structures endowed with a certain stability; they occupy a certain portion of space and lasts a certain interval of time; further, though a given object can be perceived under very different aspects, we do not hesitate to recognize it as such; the recognition of the same entity under the infinite variety of its aspects poses a problem (the classical philosophical problem of the concept), that *Gestalt* psychologists were the first to tackle in a geometric perspective accessible to scientific interpretation. Let us suppose that this problem is solved on the basis of a naive intuition that grants to the external objects an existence independent of our perception. Nevertheless we have to admit that the spectacle of the universe is an unceasing movement of birth, development, and destruction of forms. The pur-

pose of science is to predict this evolution of forms, and if possible to explain it.[1]

This problem cannot be dissociated from that of the linguistic description of phenomena. To understand the dynamics of forms in its material, efficient, and formal causes is also to understand the efficiency of the real-world descriptions in natural language, which is a central enigma whose clarification requires extremely sophisticated mathematical constructions.[2]

We see that the catastrophist point of view is distinct from the classical scientific one. For the former, the description of perceptual experience is neither superfluous nor illusory. Its possibility is rooted in the very objectivity of phenomena.[3] To draw a parallel, one could say that just as quantum mechanics included the fact of measurement as an integral part of physical objectivity, CT seeks to include the fact of description as an integral part of objectivity, to introduce in the mathematization of reality a complementarity between world and language, and thus to constitute afresh the very concept of 'objectivity'.

5.2 *Four guiding principles*

René Thom's arguments are often distinctly phenomenological in their style and content. We can discern in them four guiding principles.

5.2.1 *Phenomenological abduction*

The idea is the following. In classical physics, one proceeds forwards first from general principles to dynamics, and then from dynamics to the observed phenomenology. Here, on the contrary, one begins with phenomenology and tries to go backwards to *constraints* on the generative dynamics.

1 Thom, 1972a: 17.
2 See Thom, 1980d, Chapter II.
3 See, section 2.4.

> The method [...] essentially consists in accepting *a priori* the existence of a differential model underlying the process investigated and, without explicitly knowing this model, to deduce only from the supposition of its existence, conclusions about the nature of the singularities of the process. From this fact, certain consequences of a local and qualitative sort could be obtained from the hypothetical existence of the model.[1]

This principle *reverses* the usual top-down order of physical deduction. But it is not an induction either. It is a sort of backward bottom-up deduction. What Peirce called an *abduction*.

> Our central idea is that the processes of morphogenesis are in fact determined by an underlying dynamics, which in general would be impossible to make explicit. [...] One can, to some extent, classify and predict the singularities of the system's morphogenesis, even without knowing, either the underlying dynamics, or the dynamics of macroscopic evolution. [...] In fact, in most cases, one will have to proceed in the reverse direction: *from a macroscopic examination of the morphogenesis of a process, from the local or global investigation of its singularities, one will try to go back to the dynamics that generates it.*[2]

These models which aim at

> an analogical classification of the dynamical situation supposed to generate the experimental morphology[3]

reintegrate appearances within objective reality. They interpret appearances as the phenomenal 'externalization' of the 'interiority' of systemic 'black boxes.'

5.2.2 *Ontological neutrality and phenomenological reduction*

According to Thom,

> One of the essential features of the local modeling method suggested here is that it assumes nothing regarding the ultimate nature of reality; even if this reality should be of a complexity beyond description, only some of its aspects

1 Thom, 1975: 23–24.
2 Thom, 1980a: 101.
3 Thom, 1972a: 20.

> finally enter in the macroscopic description, namely, those which determine the 'observable' parameters of the system.[1]

The idea was therefore to bracket the fine-grained physics underlying natural phenomena and to retain only their salient qualitative morphologies. It was quite similar to Husserl's epoche (phenomenological reduction). As far as I know, it was the first time that such a phenomenological principle was introduced in natural sciences. We will see later that it is the bridge linking the physical foundations of phenomena with their semiotic form.

5.2.3 *The independence from the substratum*

CT aspires to be

> a theory of morphogenesis *in abstracto*, purely geometric, independent of the substratum of forms and of the nature of the forces which create them.[2]

> In catastrophe theory, there is need for a synthetic method which, to some extent, is inherited from the old *Naturphilosophie*. In my opinion, if we observe phenomena from a distance, we notice that several morphological accidents seem to be independent of the nature of the entities they involve. The classification of these general and ubiquitous phenomena allows us to isolate 'entities' which operate locally in these dynamics, and which I refer to as *logoï*, or archetypes. In principle, these archetypes can manifest themselves on any substratum whatsoever. [...] The theory of elementary catastrophes is in some way a theory of the most general substratum, that is, of the undifferentiated substratum, and I would say, of the *materia prima* of the scholastics.[3]

This principle is certainly most astonishing:

> The main idea of our theory, namely that a certain understanding of morphogenetic processes is possible without recourse to the special properties of the substratum of forms, or to the nature of the active forces, could seem difficult to accept, especially for the experimental scientists used to cutting the flesh and who are always confronted with a reality that resists them.[4]

---

1 Ibid.: 23.
2 Ibid.: 24.
3 Thom, 1980d, Chapter III.
4 Thom, 1966, and 1980a: 10.

It asserts that for the morphological order, causality is structural and formal before being physical and material. It is justified by:

(i) the evidence that, at least locally, the morphology and the morphogenesis of phenomenological accidents are essentially *underdetermined* relative to the internal generative dynamics, and

(ii) the demonstration that they undergo drastic mathematical (geometrico-topological) constraints, so drastic as to permit, in the elementary cases, the reconstruction of a minimally complex generating dynamics.

It does not surprise anybody today, if we say, for example, that space-time geometry constrains the physics of elementary particles to the extent of largely determining it. Or that for purely topological reasons the movements of integrable Hamiltonian systems are quasi-periodic movements on invariant tori, or still that self-reproduction requires, according to Von Neumann's theorem, a structure of 'genetic code' type. Thus, we needn't be astonished by the fact that the morphological order can also be constrained by a geometric *eidos* that 'nature' is 'obliged' to realize materially.

5.2.4 *Hylemorphism*

> In a being – or object – we distinguish classically its existence, i.e., the fact that it fills a certain portion of space-time, and its essence, that is, the totality of its attributes and qualities. The materialist perspective, common in science, insists that existence precedes essence (in fact, existence implies essence); the CT model of morphogenesis counters this axiom, for it presupposes that, to a certain extent, existence is determined by the essence, the set of qualities of the being. What we have here is a return to the Aristotelian notion of hylemorphism: matter aspires to form.[1]

The general silence on these questions is largely due to the Galilean-Newtonian tradition which imposed an ontological primacy of force over form.

---

1 Thom, 1980a: 87.

> There is hardly any reason to think that force has, in principle, a deeper ontological status than form. [...] I think that, in a very general sense, the concept of form is infinitely richer and more subtle than that of force.[1]

The subtlety and the richness of the concept of form becomes particularly evident when we move from the local archetypes to their integration in stable global structures.[2] In such an integration the specific internal dynamics recovers their rights.

> Matter often imposes additional constraints of rigidity, symmetry, and certain invariance of volumes, etc., and consequently, the theory of singularities has to be modified. This modification manifests itself empirically in the specific nature of the singularities: for example, the singularities of clouds are not the same as the singularities of an iceberg, or of a rock.[3]

> Why is it that the form of clouds is not the same as that of mountains, why is the form of crystals not that of living beings? I would answer that our model aims only to classify the *local accidents* of morphogenesis that we call *elementary catastrophes*. But the macroscopic global appearance, or form in the ordinary sense of the term, comes from the aggregation of a large number of such local accidents; and the statistics of these local catastrophes, the correlations which govern their appearance in the course of a given process, are determined by the topological structure of their internal dynamics; the integration of all these accidents into a global structure will require – if we wish to pursue the application of our model – the consideration of catastrophes on spaces of dimensions much larger than the usual three. It is because of the topological richness of the internal dynamics, and their more or less integrated character, that one can ultimately explain the nearly infinite diversity of appearances in the external world, and perhaps also the fundamental distinction between life and non-life.[4]

5.3 *CT as a mathematical phenomenology*

As soon as CT legitimizes the bracketing (phenomenological reduction) of the internal generating dynamics, as soon as it can provide a precise mathematical interpretation of appearances as the expression of being, it can be applied as well to physical substrata where one can, in principle if not in fact, make explicit the internal dynamics, as

1 See, Thom, 1980d, Chapter III.
2 For an examination of the Local/Global opposition, see Petitot, 1979b.
3 Thom, 1980d, Chapter III.
4 Thom, 1972a: 24–25.

to non-physical substrata where such an explicitation is impossible. We see that CT always aims at a mathematical description of the morphological manifestation, but, depending on the case, it opts for either a phenomenological bracketing or a physicalist explicitation regarding the generating dynamics. Whence its ontological neutrality. It is compatible *both* with physicalist reductionism which causally deduces form from matter and hylemorphic idealism which ascribes form to matter.

CT is thus a *mathematical phenomenology* that works out a synthesis of the physicalist and structuralist viewpoints; what Thom calls a *geometrization of concepts* establishing a mediation between natural phenomena and signification (hence between natural sciences and semiotics). It functions in opposite directions in natural sciences and social sciences. In natural sciences, it results in the integration of a morphological phenomenology to physical objectivity. But, in social sciences, it naturalizes semiotic structures.

Actually, I believe that the transcendental relevance of CT is that its 'step back' towards appearance and manifestation is at the same time a naturalization of meaning.

> In social sciences the use of natural language and of purely conceptual thinking leads to a way of reasoning which is often extremely intricate and subtle. […] If we are able to translate such reasoning into a purely geometric (topological) picture, then we may ensure to a large extent the objectivity of this thinking; by using the 'distanciation' effect of geometric representation, we can break the hermeneutic circle which has kept imprisoned so many of social science thinkers.[1]

> One of the biggest difficulties in Semantics is that, when we wish to analyze meaning, […] we are *in* meaning. […] That is why, to succeed in doing an objective and scientific analysis of meaning, we should be able to be distanced from it. […] In this lies the great interest of a geometrization of meaning. To the extent we can geometrize the processes bearing significations by rendering them inert, we can submit them to a combinatorics which falls outside the traditional semantic categories. It is precisely this type of analysis which is made possible by the geometrization associated with Catastrophe theory. […] Geometrization of the semiotic processes is extremely interesting because it is able to break the 'semiotic circularity'.[2]

---

1 Thom, 1980b.

2 Thom, 1980d, Chapter IV.

Thus catastrophist hermeneutics, far from eliminating meaning in a formalist manner, attempts to reduce its subjectivity by substituting

> semantic intuition which is of a directly subjective nature, with geometric intuition which spatializes its object, and distanciates it from the thinking subject.[1]

Of course, this substitution is possible only by considerably extending geometric intuition. In this sense CT is a 'language' which is 'formal' in an entirely new sense. It is not a symbolic, but a topologico-geometrical language, whose semantics is geometrized and whose syntax is constructed locally from the most simple and archetypal events and interactions. Just as for the morphological order the integration of local accidents into a global structure is a central problem, so for this language, contrary to the formal languages, integration of local syntactic structures, iteration and recursion, in short, generativity, constitute a central problem, still completely unexplored and which should become the goal of a 'dynamical topology'.

5.4 *Critique of logicism*

It is then understandable that the catastrophist point of view levels constant objections against the purely symbolic points of view. Its critique is twofold.

5.4.1 *Extensionality/Intensionality*

Firstly, though highly relevant in mathematics, the logicist-formalist conception is fundamentally inappropriate to the study of natural languages because:

(i) linguistic concepts, as opposed to mathematical ones are vague and 'non-constructed' concepts;

(ii) grammatical recursivity is so poor that it is not a recursivity in the logical sense of the term.

1 Thom, 1980a: 123.

> Frege [...] stands in opposition to Russell and Hilbert [...]. He always wanted the axioms to be *true*, and not empty postulates. His logic was basically intensional and not extensional. It is only by invoking the 'principle of extensionality': 'the intension of a concept determines its extension and vice versa', that he was able to establish a logic of a combinatorial and formal type. Now, closer observation shows that the extension of a concept in natural language is a 'fuzzy' set whose limits can never be defined – supposing that they exist independently of the idiolect of any individual speaker. This means that every extensional logic is fundamentally inadequate to describe the mode of reasoning specific to natural language. Hence the basic impossibility of reducing a theory of natural language to a Boolean or Fregean type of logic. A logic that would account for natural reasoning is necessarily intensional and it is obtained necessarily via a theory of the concept seen as an intensional entity. How can we, faced with such massive evidence, still hold on to dogmatic anti-psychologism?[1]

> Any true logic has to be intensional and cannot be extensional unless it deals with concepts of an artificial kind, whose extension can be generated by a constructive procedure.[2]

Whatever be the advances made by Kripke, Hintikka, and others in their interpretations of intensional logics in terms of possible worlds, the central question of a *qualitative* logic of the concept is still widely open.

5.4.2 *Cinematics/Dynamics*

The second criticism leveled against the formalist-logicist points of view no longer concerns their inadequacy to the structure of natural languages, but their ability to model phenomena adequately. When we model a real system by a formal system *P*, we assume that

> every state *A* of the phenomenological process under consideration can be parametrized by a system of propositions *a* of the formal system *P* [and that] if, in course of time, the state *A* is transformed into the state *B*, *B* can be parametrized by a set *b* of the system *P* such that *b* can be formally deduced from *a* in *P*.[3]

1 Letter of R. Thom to G. G. Granger (July 28, 1979).
2 Thom, 1980b.
3 Thom, 1972a: 18.

In other words, we assume that we can interpret temporal succession in terms of logical implication. But,

> every model consists *a priori* of two components: a cinematic component whose role is to parametrize the forms or the states of the process, and a *dynamic* component whose role is to describe the temporal evolution between the forms.[1]

The logicist conception implicitly postulates that *a formal cinematics can stand for a dynamics*. Such a dogma has evidently many advantages:

> An axiomatic or combinatorial type of description is very easy; deduction is formalized, and theoretically mechanizable.[2]

But it remains nevertheless a fundamentally erroneous conception, for *'no dynamics is possible'* in that context. Hence the necessity of introducing dynamical models for the topological understructures: in order to overcome the triviality of their formal cinematics, structural formalizations should include underlying dynamics providing a morphological realization of their combinatorics and their surface axiomatic rules.

5.5 *Centrifugal dynamics*

At the beginning of Chapter 10 of *Structural Stability and Morphogenesis*, René Thom comments on the following passage from Uexküll's *The Theory of Meaning* which sums up the problem:

> Any machine, for example, a pocket-watch, is always constructed centripetally, that is, all the parts of the watch – hands, spring, and wheels – must always be made first before being mounted on a common frame. On the contrary, the growth of an animal, such as the triton, always takes place in a centrifugal manner, starting from its germ; the *gastrula* appears first, followed by new buds which develop into differentiated organs. In both cases, there is a construction-plan (a design); the watch-plan proceeds centripetally, and the

1 Ibid.: 19.
2 Ibid.: 37.

> triton-plan centrifugally. Depending upon the plan, the parts are assembled according to completely opposite principles.

Thom says:

> I do not think that there can be a better way than that description of the physiologist Uexküll for characterizing the essential difference that separates the vital dynamics from anthropomorphic constructions with which it is often compared. It is not that the similarities between vital mechanics and certain aspects of human technical innovation (automata, electronic computers, etc.) are without any value: but these comparisons are valid only for mounted partial mechanisms which are in full functional activity: they can in no case be applied to the global structure of living beings, nor to their epigenesis and their physiological maturation.[1]

The major difficulty with organizational models is to be *compatible with the ontogenesis* of the phenomena they are modeling, that is to say, *to implement the formal cinematics in the underlying generating dynamics*. For structural models, the difficulty concerns the genesis of deep structures.

> The so-called 'deep structures' [...] are not really so deep! They are only equivalence classes of surface structures, obtained by means of relatively trivial transformations. Instead, for me, it would be far more interesting to reveal the generating dynamics of deep structures, [...] just as in biology it would be interesting to reveal the dynamical processes which generate the biochemical morphologies that the biologist studies. But, alas, it lacks an adequate conceptual equipment.[2]

Due to this lack of any dynamical perspective on deep structures, formal linguistics tends to equate deep structures with their formal cinematics and to solve the question of their genesis using unverifiable innatist hypotheses.

The same can be said about the conception of structural paradigms. Every model of paradigmatic structures should formalize the dynamical processes of differentiation of the semantic substrata into positional values, i.e., *the genesis of the thresholds* which categorize and discretize them. The logico-combinatorial structuralism postulates

1 Ibid.: 207.
2 Thom, 1980d, chap. IV.

that the thresholds are constitutive while at the same time treating them formally as already constituted.

> If we wish to understand how a threshold appears, we are obliged to adopt an ontogenetic as well as a diachronic perspective which explains the genesis of the threshold. But if we wish to explain the genesis of a threshold, we are quite automatically led to consider a situation of the catastrophe type. Evidently, we can also be not interested in the genesis of the threshold and simply wish to understand how an already constituted threshold functions.[1]

5.6 Phenomena as morphologies

Thus we see that the catastrophist strategy depends on a redefinition of the primitive 'phenomenon', which would be at the same time phenomenologically faithful, compatible with physics, and valid for non-physical domains.

This renewed definition view a phenomenon as a morphology, i.e., as a system of qualitative discontinuities on a substratum space. This is an *a priori* which, in René Thom's ontology, plays the same role as the *a priori* of spatial extension in classical rationalism. Discontinuity is a pure intuition. Beyond its evident empirical realism, it possesses a transcendental ideality by which it conditions the appearances of phenomena. Discontinuities are inherent to objectivity. But their reality is also perceptual since

> the discontinuities of morphologies constitute the most (perceptually) salient and the most stable elements.[2]

In conceiving them as 'subjective' in the transcendental sense, we are following Kant's practice where the invariants of perception are posited as pure intuitions.

If every phenomenon is, in its phenomenological appearance *and* its physical objectivity, a system of discontinuities, then the primary task of scientific explanation is, as we have seen, to mathematically model these systems so as to account not only for their physical origins but also for their describability in natural language.

---

1 Ibid., chap. III.
2 Ibid.

For this, we shall analyse the morphologies into aggregates of stable local accidents (chreodes), and we shall seek:

(i) local dynamical models for the chreodes, and

(ii) global dynamical models of integration and combination of chreodes into global structures.

If such a research program has been formulated only recently,[#] it is because every discontinuity is a *critical phenomenon* – a symmetry breaking of the substratum homogeneity – induced by a *singularity* of the underlying dynamics. It depended therefore on mathematical and physical breakthroughs in the conceptual and technical treatment of singularities, bifurcations, and structural stability of non-linear dynamical systems.

The relation between CT as a modeling procedure and CT as a morphological language is well summarized at the end of *Structural Stability and Morphogenesis*.

> 1. Every object, every physical form, can be represented by an *attractor* C of a dynamical system in a space *M* of internal variables.
> 2. Such an object is stable and can be perceived only if the corresponding attractor is *structurally stable.*
> 3. Every creation or destruction of forms, every morphogenesis, can be described as the disappearance of the attractors representing the initial forms and their replacement, through capture, by the attractors representing the final forms. This process, called *catastrophe*, can be described on a space *P* of *external variables.*
> 4. Every structurally stable morphological process is described as a (system of) catastrophe(s) on *P* that is (are) structurally stable.
> 5. Every natural process can be broken down into structurally stable parts, or *chreodes.* The set of chreodes and the multidimensional syntax which orders their respective positions constitute a *semantic model.*
> 6. If we consider a chreode C as a word in this multidimensional language, the signification of this word is nothing but the global topology of the associated attractors and the catastrophes that they undergo. Especially, for a given attractor, the signification is defined by the geometry of its domain of existence in *P* and the topology of the regulation catastrophes which bound this domain.[1]

---

i.e., in the seventies.

1 Thom, 1972a: 321.

This interplay of physics, perception and semiotics, proceeding from the notion of discontinuity as pure intuition, its treatment within singularity theory, and the principle of structural stability, constitute the phenomenological essence of CT as a synthesis of the physical and structural realms.

5.7 *The locality principle*

Though in part hermeneutic, CT is nevertheless properly scientific to the extent that it replaces the conceptual 'magic' with a geometrization that satisfies the criteria of *locality*. The locality requirement is fundamental in physics where major breakthroughs have consisted in localizing classical theories (electromagnetism with Maxwell, gravitation with Einstein). With the project of geometrizing concepts, CT extends this scientific imperative to non-physical descriptive-conceptual theories, and in particular to biology and linguistics.

But if the locality principle happens to be one of the primary criteria of scientificity, the essence of objectivity will basically depend on the possible extensions from local to global. Now, Thom insisted that, in a very general way, physics relies on the specific extension from local to global which is called analytic continuation in complex analysis.

> Pragmatically efficient and predictive mathematical models imply the analyticity of the functions they involve, and of their solutions for temporal evolution. Consequently, this implies that the 'substratum' space on which one works must be provided with a natural analytical structure. Only the *analytic continuation* would allow for the extension from local to non-local that characterizes action.[1]

In fundamental physics, the ground space-time is endowed (at least locally on the cosmological scale) with a natural analytic structure. Furthermore, *all the other spaces are explicitly derived from this ground space-time and inherit, in one way or the other, its analytic properties.*

> In fundamental physics, the internal spaces which must be introduced for describing the physical entities can be directly related to space-time or to its

1 Thom, 1980a: 116.

> equivalence groups by well-defined mathematical constructions. *Nothing more is needed for explaining the main fundamental laws and their analytic character.*[1]

The fundamental physical laws would express the analytic properties of 'the 'regulation' of space-time vis-a-vis the accidents it undergoes.'[2]

If Thom emphasizes this point so much, it is because it characterizes physical objectivity within an *extended* rational framework and therefore allows to define *alternative* types of objectivity. Actually, the notion of singularity represents another grand mathematical procedure – alternative to analytic continuation – for the extension from local to global. Singularities can unfold in spaces endowed with a 'weak' geometrical structure. That is why,

> instead of the global regulation of space-time, we can envisage these local qualitative regulations which give birth to the typical forms (animate or inanimate) listed under recognizable (and identifiable) individuals.[3]

Thus, structural 'physics' which conflates a phenomenological revival in natural sciences and a physicalist objectivization in structural semiotics is still a 'physics'. But, it is a 'physics' founded on an altogether different procedure of extension from local to global and which, as a consequence, does not satisfy any one of the normative scientific criteria prevalent in fundamental physics. It is an alternative 'physics' which transforms the morphological order into a new order of objective legality.

5.8 *Mathematics and reality*

In attempting to geometrize concepts, CT seeks to overcome the positivist conception of rationality. It reintroduces the primacy of the theoretical dimension and reopens the question of the role of mathematics in the constitution of objectivity.

---

1 Ibid.: 118 (our emphasis).
2 Ibid.
3 Ibid.

The concordance, often observed in several disciplines relating to animate and inanimate world, between an empirical morphology and a mathematical structure brings up a classical problem of epistemology. We can address it with three types of responses:

1. The first attributes this agreement to a 'pre-established harmony' between mathematics and reality. This is the Platonic (more exactly, Pythagorean) response: God always employs geometry.
2. The second attributes the appearance of the mathematical structure to a phenomenon of local equilibrium, or as it is said in Mechanics, to the solution of a problem of extremality.
3. The third – which I advocate – attributes the appearance of the structure (and the morphological repetitions that it gives rise to) to a hypothesis of *genericity*: in all circumstances, nature realizes *the least complex* local morphology compatible with initial local data.

The first response is pure metaphysics. Only the second can be considered strictly scientific, because it can be sometimes submitted to quantitative testing. [...] The third response is mid-way between science and metaphysics. [...] The second and third viewpoints are moreover not incompatible. [...] The advantage of the third point of view lies in not taking sides at once on the question of determinism in the evolution of structures. In the third perspective, determinism is less a given *a priori* than a conquest.[1]

1 Ibid.: 143–144.

CHAPTER II

Categorical Perception and Topological Syntax

A double application of morphodynamical models to the double articulation of language

In this chapter we will see how and why the catastrophist point of view is required in two key domains of structuralism, namely, phonology and structural syntax, corresponding respectively to the second and first articulations of language. We begin with phonology (Sec. 1) since it is, we recall, at the foundation of modern structuralism. The analysis of the fundamental phenomenon of categorical perception (Sec. 1.1 and 1.2) – already referred to in I.2.5 – will further provide a natural opportunity to sketch the principles of catastrophist models (Sec. 1.3).[1] We will then present some reflections on structural syntax. After describing the actantial schematization for different conceptions of grammar (Sec. 2, 3, 4) and criticizing the formalist point of view of transformational-generative grammars (Sec. 2.2, 4.1), we will show the close relationship that exists between catastrophist schematization and case grammars (Sec. 5). Then we will go on to identify the topologico-dynamical conception of deep cases with a renewed version of the localist hypothesis (Sec. 6).

1 We will be rather concise concerning this theme, for we have treated it extensively in our book on phonology: Petitot, 1985.

1 Phonological categorizations as critical phenomena

Let us elaborate on the generalities already presented in Sec. I.2.5.

1.1 Categorical perception

1.1.1 Definition

Discovered by A. Liberman in 1957, categorical perception is contrasted with continuous perception. Let us consider a 'continuum' of stimuli ranging from an initial syllable $S_1 = C_1V$ to a final syllable $S_2 = C_2V$ with the same vowel V, the consonants C_1 and C_2 (in general stops) differing only by a single acoustic cue (e.g. voicing as in [ba]/[pa], [du]/[tu], etc., or place of articulation as in [bo]/[go], [pi]/[ti], etc.). This ideal 'continuum' is in fact concretely a discrete series of N stimuli (where N is in general between 10 and 20) of which the first and the last are natural (i.e., articulatorily producible) and the others synthetic. With respect to this experimental material, a group of subjects are submitted to tests of identification and discrimination.

The results show that there is *no intra-categorical discrimination*: subjects discriminate two neighboring stimuli n and n+1 only if they are on either side of an interface separating two adjacent categories. In other words, and contrary to what happens in cases of continuous perception such as color perception, discrimination is subordinated to identification, that is, it takes place on an absolute and not a relative basis (see fig. 1).

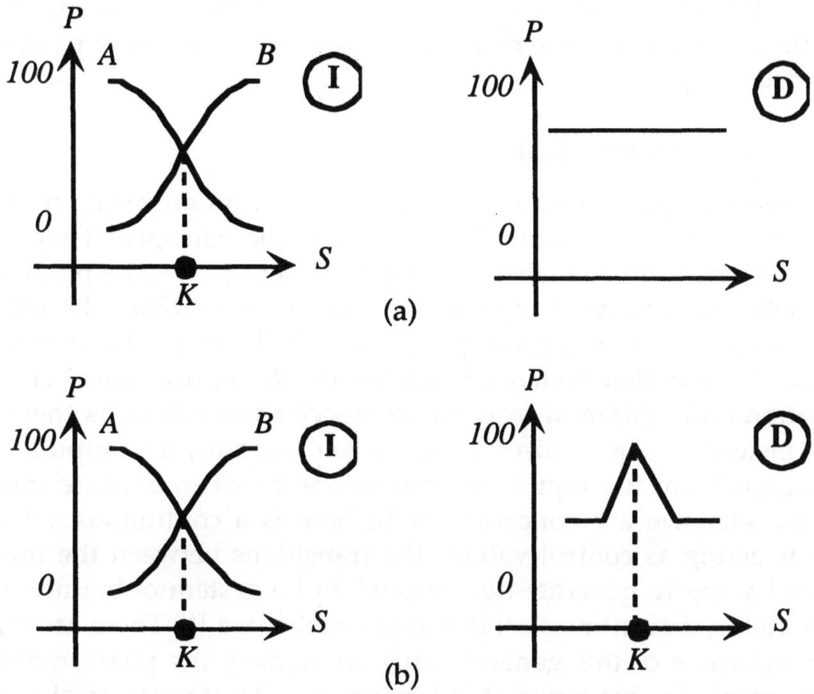

Fig. 1. (*P* = %, *S* = Stimuli, **I** = Identification, **D** = Discrimination, *K* = Boundary). (a) Continuous perception. (b) Categorical perception.

As M. Studdert-Kennedy and A. Liberman pointed out,

> categorical perception refers to a mode by which stimuli are responded to, and can only be responded to, in absolute terms.[1]

1.1.2 *Function*

The functional importance of categorical perception is evident. It is the absence of intra-categorical discrimination which ensures perceptual discretization, and thus enables the audio-acoustic flow to be the base for the phonological code. This discretization essentially concerns the consonants (and especially stops), that is, the phonemes which are strongly encoded in the flow (the perception of vowels and fricatives is more continuous than categorical). The encoded pho-

1 Studdert-Kennedy *et al.*, 1970.

nemes are categorical as immediately given to perception and probably there exists a specific mode (a 'speech mode') for their processing and decoding.

1.1.3 *General abstract situation*

The phonetic phenomena of categorical perception result from the manner in which the acoustic cues control the percepts.[1] They correspond to the following abstract situation. Let $u_1,\ldots, u_n$ be parameters (acoustic cues) varying over a space W and controling the internal states of a 'black box' S (perceptual system). What is to be understood is how a controlled system can categorize its control space. This is a situation quite different from those described in automata theory. In fact, instead of sets of discrete inputs and outputs, the outputs being produced from the inputs via transitions between discrete internal states, what we are concerned with here is a continuous set W of inputs acting as control values, the transitions between the internal states having to generate not outputs, but a system of boundaries K (thresholds, discontinuities) in the external space W. There are typical physical cases of this general situation, namely the phase-transition phenomena. In this sense, it is legitimate to treat categorical perception as an induction of 'phase diagrams' in the spaces of acoustic cues controlling the percepts.

1.1.4 *Examples*

In the late sixties and the early seventies, a number of crucial experiments were conducted on categorical perception. The boundaries K induced on the VOT axis [2] by the identification of basic pairs of stops [b]/[p], [d]/[t] and [g]/[k] were studied intensively. Pioneering experiments were conducted in 1970 by Lisker and Abramson who analyzed the variation of K as a function of the place of articulation (see fig. 2).[3]

---

1 See Liberman *et al.*, 1967.
2 The VOT (voice onset time) is the acoustic cue for voicing.
3 See Lisker, Abramson, 1970.

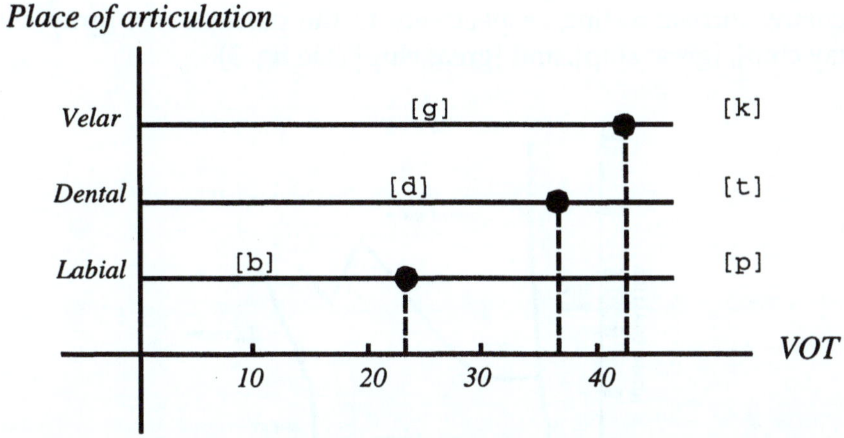

Fig. 2. Lisker's and Abramson's experiments are insufficient: 3 points don't permit to reconstruct a 2-dimensional categorization.

But these experiments are still quite insufficient. In fact, since place of articulation (as well as voicing) depends on *several* continuous acoustic cues (e.g. the frequency of the plosive burst and the transition of the second formant, cf. P. Delattre's locus theory), the boundary system K induced by categorical perception categorizes a *multidimensional* external space W of dimension r. Now, for classifying and discriminating the percepts controlled by W, it is necessary to decompose W into domains (categories). This requires boundaries K of codimension 1 (i.e., of dimension r–1). Further, the fundamental information is the geometric one provided by the morphology of K. Now, as is clear from figure 2, Lisker's and Abramson's results do not permit to reconstruct a morphology of codimension 1 (i.e., of dimension 2–1 = 1) in the external space of the VOT and the place of articulation.

However there have been successful attempts at an explicit reconstruction of a phase-diagram in an acoustic control space. One example is B. Repp's experiment on English fricatives.[1] Repp takes two control parameters, a period of silence ΔS and a period of fricative noise ΔB, and analyzes their cooperation in the discrimination of fricatives and affricates. In the case of an utterance like 'Did anybody see the gray ship', the external space (ΔS, ΔB) is categorized into 4

1 Repp *et al.*, 1978.

domains corresponding respectively to the perceptions: [gray ship], [gray chip], [great ship], and [great chip] (see fig. 3).

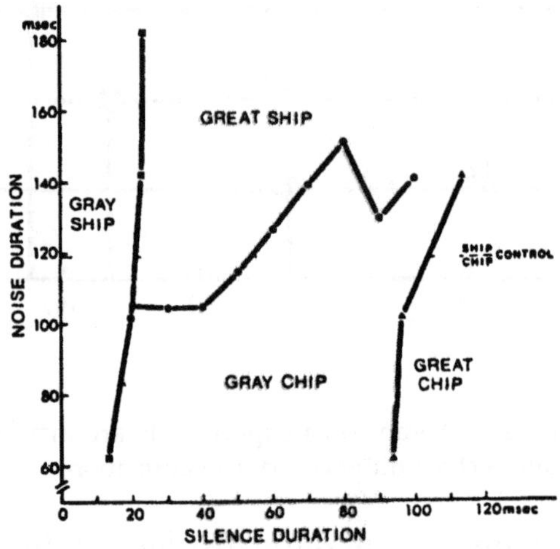

Fig. 3. Repp's phase diagram (Repp *et al.*, 1978). The boundaries divide the control space of duration of silence and duration of fricative noise in different categories.

1.1.5 Specificity

Contrary to initial assumptions, it was discovered that categorical perception is not specifically phonetic. It exists ubiquitously in the whole auditory domain. For example, the perception of musical timbres is in part categorical.[1] And also the perception of musical intervals by professional musicians possessing an 'absolute ear'.[2] The fact that professional musicians process musical information preferentially in the left hemisphere (the dominance of the right ear can be tested by experiments on dichotic listening) show there exists in that hemisphere a specific categorical mode of information processing, able to discretize (digitalize) information and transform it into a code.

There are also phenomena of categorical perception involving only temporal organization at a very abstract level. For instance J.

---

1 Cutting, Rosner, 1974.
2 Siegel, Siegel, 1977.

Mehler showed that if three beats 1, 2, 3 are distributed on an interval of 600ms, with the beat 2 situated in a variable intermediary position, the perception is categorical. It divides the stimuli in 3 classes corresponding respectively to the invariant perceptions 1–2/3, 1/2/3 and 1/2–3, the boundaries being situated at about ± 20–30ms from the central position (beat 2 at 300ms).[1]

There exist phenomena of categorical perception in the visual field too. We get a typical example with the intermittent light stimuli, where there exists a frequency threshold beyond which the stimuli are perceived as continuous ('flicker-fusion' threshold).

1.1.6 Innateness

Let us return to the categorization of the VOT axis. Experiments on languages like English or French where there are only two possibilities of voicing (voiced/unvoiced) reveal a boundary separating [d] from [t]. But a language like Thai, where there are three possibilities of voicing, reveals *two* boundaries separating respectively [d] from [t] and [t] from [t^h] (aspirated [t]). Experiments of this kind show that *phonological paradigms* result from the boundaries K induced in the spaces of acoustic cues W by categorical perception. These boundaries vary from language to language (the boundary [b]/[p] on the VOT axis is at 37ms for English and 5ms for French), confirming the relativity of phonological categories. But it seems that these systems K forming part of the *steady state* of a language proceed from an *innate initial state* K_0.[2] In fact, cross-linguistic experiments on preverbal children seem to show that there exists a universal (genetically determined) categorization of the VOT axis defined by two boundaries, the first at about –30, –20ms, and the second at about +20, +30ms (see fig. 4).

1 Mehler, Bertoncini, 1980.

2 For the debate concerning innateness, see the Chomsky/Piaget confrontation: TLTA, 1979.

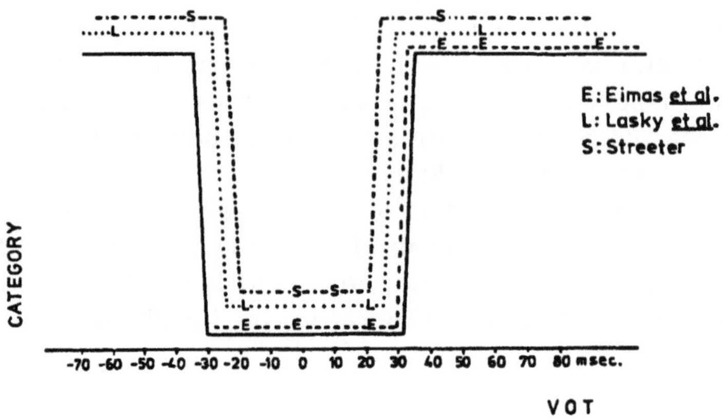

Fig. 4. The universal VOT categorization in pre-verbal children. From Mehler, Bertoncini, 1980. Lasky, 1975 concerns Spanish babies, Streeter, 1976 Kikuyu babies.

The existence of an innate sensorial component, though challenging the old idea of the *tabula rasa*, is *a priori* necessary for the acquisition of language. For, as P.D. Eimas [1] emphasized, for a child to learn his first language, he must be able:

(i) to discriminate between small differences in the acoustic signals;

(ii) to categorize a continuous sequence of acoustic values;

(iii) to grasp the intrinsic organization of the basic units, phonemes and syllables;

(iv) to extract invariants of perception despite the variations of the signals;

(v) to process in a context-dependent way the acoustic information which is critical for phonetic distinctions.

1 Eimas, 1980.

1.2 *Conflicting Interpretations*

There are various interpretations of categorical perception. They differ in relation to classical oppositions like sensorial/cognitive and reductionist/structural. These conflicting positions are of special epistemological interest for they demonstrate that without a deep theoretical investigation, it is almost impossible to choose between them experimentally.

1.2.1 *The sensorial hypothesis*

Many workers have hypothesized that categorical perception is a general property of perception, deducible from psychophysical principles. For instance, according to J. Miller,[1] it only shows the presence of a masked sensorial threshold. In all the experiments, one considers a 'continuum' associated with the variation of a cue I discretized in equal steps ΔI. Below the threshold, the variations ΔI would not be detectable (and thus not discriminated). And beyond the threshold, they would still not be detectable, according to Weber-Fechner law.

This hypothesis has been used by R. Pastore to account for categorical perception without subordinating sensory discrimination to cognitive identification. According to Pastore, categorical perception is due to the structure of peripheral neuro-perceptual processes and proceeds from a limitation, internal or external, which is better defined than the limen of differentiation of the continuum under investigation. The limitation is internal when it corresponds to a masked threshold. It is external when it introduces a reference stimulus with which the other stimuli are being compared.[2]

1.2.2 *The reductionist hypothesis of feature-detectors*

The experiments on preverbal children have challenged the classical motor theories of phonetic perception which consider it as an articulatorily finalized process, resulting from a 'mental reconstruction' of

---

1 Miller *et al.*, 1976.
2 Pastore *et al.*, 1977.

the articulatory process.[1] In fact, since infants begin to perceive before they can articulate, we are led to suppose that there is a genetic determination of the feedback from perception onto articulation. Now

> to attribute this knowledge to the infant's biological endowment would seem to extend considerably the cognitive competencies that we are willing to impute to genetically determined factors.[2]

Further, some experiments on dichotic listening seem to suggest that perception takes place by recombining, at the central cortical level, the distinctive features extracted from acoustic signals at the peripheral level. Since their discovery by Eimas and Corbitt in 1973, we have at hand a certain number of convergent experiments on the phenomena of *selective adaptation*.[3]

In these experiments, one of the extreme stimuli from a continuum *W* (for example ([ba] – [pa]) is presented repeatedly to subjects (adaptation). After adaptation, the boundary *K* categorizing *W*, relative to the situation before adaptation, shifts towards the adapting stimulus. Moreover, adaptation can be induced not only by a stimulus of the series tested, but by a stimulus sharing with it only one distinctive feature (selectivity of adaptation).

These results prompted Eimas and Corbitt to come up with the hypothesis of *feature-detectors*, which are neuro-sensory receptors responding selectively to well-defined domains of values of acoustic cues:

> 'Feature-detectors' can be broadly defined as organizational configurations of the sensory nervous system that are highly sensitive to certain parameters of complex stimuli.[4]

If we suppose that they are part of the initial phonetic state of the organism, that they are sensitive only to restricted domains of variation of the acoustic cues, and that their domain of response is ac-

---

1 Motor theories assume also that articulation is phonologically finalized by internalizing a space of phonetic targets (see Mac Neilage, 1979).

2 Eimas, 1974: 518.

3 Eimas, Corbitt, 1973.

4 Abbs, Sussman, 1971: 24.

quired through interaction with the environment, then we can easily explain the above experiments. It would be enough to assume that:

(i) the VOT axis is covered by the domains of two (or three) detectors;

(ii) the responses of these detectors are in competition; v

(iii) the information processing devices of the central nervous system are sensitive only to the excitation of the detector whose response is maximal.

(iv) repeated presentation of the same feature 'tires out' the corresponding detector (see fig. 5).

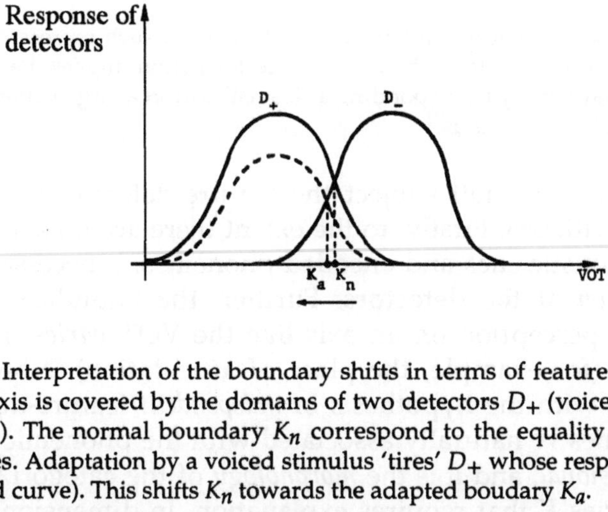

Fig. 5. Interpretation of the boundary shifts in terms of feature detectors. The VOT axis is covered by the domains of two detectors D_+ (voiced) and D_- (unvoiced). The normal boundary K_n correspond to the equality of the two responses. Adaptation by a voiced stimulus 'tires' D_+ whose response decreases (dotted curve). This shifts K_n towards the adapted boudary K_a.

If this reductionist hypothesis on feature-detectors was so influential, it is not only because it provided a simple explanation of categorical perception and of selective adaptation phenomena, but also because it introduced in phonetics a perspective already well known in visual perception since Hubel and Wiesel's works, and allowed therefore a unified neuro-physiological conception of perception. But, it is so reductionist that it should nevertheless be subjected to further theoretical scrutiny.

1.2.3 Criticism of the feature-detector hypothesis

The feature-detector hypothesis has been criticized not only by the supporters of motor theories, but also by those who uphold other reductionist theories. To begin with, as noted by R.L. Diehl, selective adaptation can be explained by other perceptual principles such as the principle of *contrast*, as per which perception privileges difference against identity. If the adapting stimulus serves as reference, the tested stimuli will be perceived much 'earlier' as different and the boundary will then be shifted towards the adaptor.[1] On the other hand, as J.S. Bryant pointed out, the effects of selective adaptation simply show that the phonological distinctive features are neurally represented.

> This does not necessarily mean that the sum of such representations is functional in perception as the feature detector notion implies. Rather, the cell or cells may simply be responding as a small part in a large pattern of neural response to the stimulus.[2]

On our part, we shall subject the feature-detector hypothesis to a three-fold criticism. Firstly, to the extent there are no simple relations between acoustic cues and encoded phonemes, it necessarily leads to a *proliferation* of the detectors. Further, the boundary induced by categorical perception on an axis like the VOT *varies* in relation to other cues (for example, the place of articulation). Now, this is incompatible with the hypothesis of *independent* feature-detectors. The spaces of cues *W* naturally associated with the phonemic Gestalts are *multidimensional*, and it is the *morphology* of the categorizing systems of boundaries *K* that requires explanation. In dimension 1, the theoretical difficulty disappears because, whatever be their generating mechanism, the *K*s are reduced to isolated points. However, it is no longer the case for *higher* dimensions where the morphology of *K* yields a simple criterion for the *refutation* of local detectors. Indeed, let a 2-dimensional space of cues *W* be covered by the response domains of a finite number of detectors whose response surfaces are 'bell-shaped' (as in the 1-dimensional case). Then K must be the projection on *W* of the intersection curves of these surfaces. But the

1 Diehl *et al.*, 1978.
2 Bryant, 1978.

morphologies thus obtained are *very different* from that encountered in critical phenomena. If we can show that the phonetic boundaries are of the latter type, then we can refute the detector hypothesis, at least its naïve version.

But the main criticism that can be levelled against this hypothesis is that it is dogmatically reductionist in comparison with the structuralist advances. Inspired by engineers in information-processing, it favors, as Bryant observed, pattern-recognition against ordinary perception and memory-stacking structures against processes. Taking for granted the reduction of the global neural dynamics to the structure of the underlying network, it adopts the computational dogma that perceptual processes are reducible to a logical 'calculus'.[1]

1.2.4 The 'foundational aporia' of phonetics

The debate about the interpretations – sensory vs. cognitive, reductionist vs. structuralist – of categorical perception is of great interest because it reveals the 'foundational aporia' of phonetics: how to understand psychophysically the phonemes as abstract, linguistically functional, units defined not by intrinsic properties but by a network of differences?

To answer we have to bring together and synthesize two completely different theoretical categorialities:

(i) in phonology, a *structural* categoriality (Saussurian, Jakobsonian, Hjelmslevian) involving positional identities, differences, discrimination, reciprocal determination, paradigmatic systems, stratifications, etc.;

(ii) in phonetics, a *psychophysical* categoriality, involving spectral forms, deformations of these forms, controls by acoustic cues, invariance/variability, categorization, boundaries, etc.

The problem is evidently to *unify* these two complementary categorialities. This is impossible if the former is interpreted in a logico-combinatorial way and the latter in a physicalist-reductionist way.

The feature-detector hypothesis is a 'dogmatic' solution to this antinomy, where the structural categoriality is eliminated and substituted by a psychophysical categoriality. However, this is theoreti-

1 See for instance, Miller, Johnson-Laird, 1976.

cally misleading for the whole problem is precisely to constitute (and not to deny a priori) the objectivity of structural categoriality. We face here a deep *mathematical* problem of schematization of conceptual categories.

Therefore, if it is legitimate to postulate that

> at all levels of linguistic hierarchy and for all categories of linguistic elements, we must suppose an abstract and functional aspect, described uniquely in relational terms, and a substantial aspect which is described, according to the choice and the goal of the describer, in terms of articulation, acoustic structure and auditory perception.[1]

It is still necessary to understand the reciprocal dependency between the phonemic form and the phonetic substance. In the absence of such an understanding, the dogmatic reductionist conceptions, notoriously inadequate, would be countered by a rigid formalist structuralism, also totally insufficient.

1.2.5 *The a priori of categorical perception and of paradigmatic categorization*

In order to understand the reciprocal dependency between the form, the substance, and the matter of expression, we should be able to define the mathematical content of the abstract situation presented in 1.1.3 and show:

(i) that it allows to schematize in a unified way the structural and the psychophysical categorialities, and

(ii) that this schematization can be specialized into models that fit the experimental data.

In other words, before attempting to work out models, we should be able to mathematically formulate the *a priori* of categorical perception as a special case of paradigmatic categorization.

The leading idea, already outlined in I.3.3 is as follows: *the concept of categorization realizes the synthesis of the (old) concept of taxonomy with the (new) concept of control.* We are given a space W of acoustic cues (supposed to be reduced to some simple parameters) $u_1,\ldots, u_r$. A point $w = (u_1,\ldots, u_r) \in W$ is an input stimulus for a per-

1 Malmberg, 1974: 210.

ceptual 'black box' S. In S, a global dynamical process X (supposed to be reducible to neurophysiological mechanisms) defines the internal states $A, B, C,\ldots$ (acoustic percepts) and is controlled by W. Furthermore, given a $w \in W$, a device I selects the *actual* internal state, the other internal states remaining virtual.

The general situation is thus characterized by:

(i) a field X_w of dynamical processes controlled by W, that is, by a map $\sigma : W \rightarrow \mathcal{X}$ associating to each point $w \in W$ with a 'point' X_w in the generalized space (the functional space) $\mathcal{X}$ consisting of all the possible internal processes;

(ii) the device of selection (or, actualization) I.

In order to understand how such a system $S = (W, \mathcal{X}, \sigma, I)$ can categorize its control space, it is sufficient to assume that perception is not determined by the exact form of the actual internal state A, but only by its *qualitative type* $\tau(A)$. $\tau(A)$ will in general be defined by the action of a group G on the functional space $\mathcal{X}$. If X is an element of $\mathcal{X}$, its orbit X^* under the action of G is constituted by qualitatively equivalent processes defining internal states of the same qualitative type. One can try to characterize $\tau(A)$ by the values of a system of invariants $\tau_1,\ldots, \tau_k$, that is, by 'properties' of the associated percepts (phonemes).

On the space $\mathcal{X}$, there will be in general a natural topology $\mathcal{T}$ defining the neighborhoods of the processes $X \in \mathcal{X}$. Now the double existence of such a topology and of the action of G suffices to define the *structural stability* of the elements of $\mathcal{X}$. If $X \in \mathcal{X}$, we say that X is structurally stable if every $Y \in \mathcal{X}$ sufficiently close to X is G-equivalent to X. Let $K_{\mathcal{X}}$ be the subset of $\mathcal{X}$ consisting of all structurally *unstable* processes. $K_{\mathcal{X}}$ acts as a discriminant morphology, intrinsic to $\mathcal{X}$, *which categorizes* $\mathcal{X}$ *and classifies the qualitative types of its stable elements*. In other words, every space $\mathcal{X}$ of processes (and more generally, every space of forms) where the notions of deformation and of qualitative type are definable, is *naturally categorized* by a subset $K_{\mathcal{X}}$, referred to as its *catastrophic set*, and which *geometrizes* the classification of its elements. Once mathematically interpreted in terms of actions of groups on generalized spaces (and no longer in a simply set-theoretic manner), the classical concept of taxonomy reveals a 'supplement' of geometry which provides the *a priori* of paradigmatic categorization.

Let $\sigma : W \rightarrow \mathcal{X}$ be the field expressing the control of the system S by the control space W. σ embeds W in $\mathcal{X}$ and we can suppose that the embedding is structurally stable, which imposes drastic constraints on its complexity. Now, let $K' = \sigma^{-1}(K_{\mathcal{X}} \cap \sigma(W))$ be the σ-preimage of $K_{\mathcal{X}}$ in W. The categorization K of W induced by S is *deducible* from K' whence the selection device I is known. In fact the dynamical origin of K is the following. Let A_w be the actual state selected by I at $w \in W$. When w varies in W, A_w varies, but if it is structurally stable its qualitative type remains constant, which implies the invariance of the associated percept and of the properties $\tau_1, \ldots, \tau_k$. However, there generally will be *critical* values of w for which A_w enters into conflict (relative to I) with another internal state B_w and is supplanted by it. K consists of these critical values controlling *catastrophic transitions* of the actual state. Now, these catastrophes will be correlated, in a manner governed by I, either with an internal destabilization of A_w (bifurcation catastrophes) or with a conflict between A_w and B_w (conflict catastrophes). But for a process X, either a destabilization of an internal state, or a conflict between two internal states are causes of instability and a control w will therefore be critical and belong to K if and only if the situation at w is correlated via I with a situation belonging to K'. In this sense K is deducible from K'.

So, the *a priori* of paradigmatic categorization is: *categorization is the trace on the control space of the instabilities and conflicts of the internal states it controls.*

1.2.6 Elements of catastrophist phonology

The characterization of the catastrophic *a priori* of categorization, allows us to formulate the elements of a catastrophist phonology. We sketch below some of its salient features.[1]

---

1 For mathematical precisions on singularity theory, qualitative dynamics (global analysis) and CT, see e.g. Abraham, 1972, Chenciner, 1973, 1980, 1984, Goulbitsky-Guillemin, 1973, Lu, 19976, Poston-Stewart, 1978, Saunders, 1980, Zeeman, 1977.

a. Discontinuous internal features and continuous external features

Given that the cues $(u_1,\ldots, u_r)$ control properties of percepts $(\tau_1,\ldots, \tau_k)$ which are invariants of their qualitative type, the notion of distinctive features must be revised. It is necessary to distinguish between the *external* features that are the cues and the *internal* features that are the qualitative invariants. The former vary continuously, while the latter, on the contrary, vary discretely according to the two main types of opposition identified by Roman Jakobson:

(i) the *qualitative* oppositions, corresponding to the conflict catastrophes (competition of two invariants), and

(ii) the *privative* oppositions, corresponding to the bifurcation catastrophes (presence/absence of a single invariant).

For example in the case of voicing, the external feature is provided by the VOT cue categorized via categorical perception, and the internal feature by the privative phonological opposition voiced/unvoiced.

In this new theoretical framework, we become able to overcome the antinomy between continuity and discontinuity, and thus *legitimize the Jakobsonian binarism on the very basis of continuous cue variation*! This solves one of the main difficulties in phonetics:

> One of the major difficulties [...] is in relating the essentially continuous nature of speech with the essentially discontinuous nature of linguistic description.[1]

But we still have two more issues. Firstly, the external features *do not generally coincide with the cues*. Let W be a space of cues $(u_1,\ldots, u_r)$. The essential information is provided by the discriminant morphology K (the catastrophic set) induced in W by the perceptual 'black box'. However, there is no *a priori* reason why the coordinate frame $R = (u_1,\ldots, u_r)$ of W should be geometrically 'adapted' to K. The position of K relative to R affords the fundamental data about the non-independence (the *integration*) of the cues. So, it is natural to look for a coordinate frame R_K of W adapted to K. It is this frame to which the external features correspond. This explains why these features, though continuous, can be discretized and reduced to +/– oppositions (which are not to be confused with qualitative or privative

1 Ladefoged, 1972: 276. See also Massaro, 1972.

oppositions describing the *internal* features). Indeed, the question is to locate not so much the points of W as the domains of W that K differentiates, classifies and positions one with respect to the other. In general, this can be done using discrete information. Let us assume for example that W has dimension 2 (cues x and y) and that K consists of two curves γ_1 and γ_2, which intersect transversally at w_0. K divides W into four domains. Locally at w_0, an adapted frame is defined by the corresponding tangents T_1 and T_2 to γ_1 and γ_2 at w_0. Each domain becomes coded by a half-tangent (+ or –) to γ_1 and a half-tangent (+ or –) to γ_2 (see fig. 6).

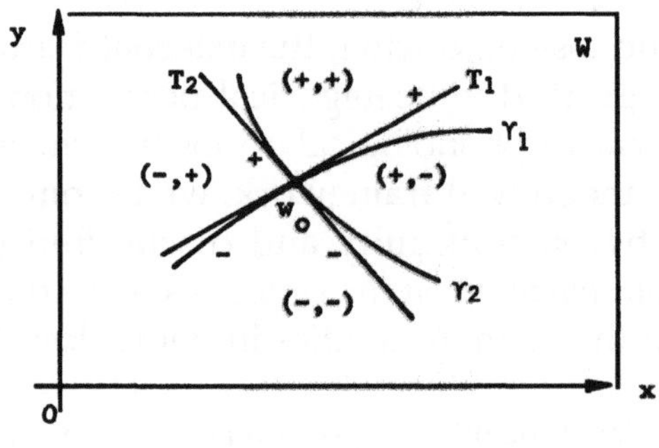

Fig. 6. Discretization of a frame adapted to $K=\gamma_1 \cup \gamma_2$. Each domain of W defined by K is coded by a pair (±,±).

It follows that *as far as the external features are concerned,* Jakobsonian binarism (valid for the internal features) should be interpreted as an hypothesis about *the type of local complexity of phonetic boundaries*: they are sufficiently simple for their adapted frame to be discretized in a binary way. But it is *not* the case in general. Due to complex integration of non independent cues, the combinatorics of distinctive features is *constrained,* and the local structure of K may drastically differ from that described in fig. 6 (which is a case of free combination).

The second reason why the *elementary* catastrophic models cannot be exact models for categorical perception is that the phonetic systems have *two* levels of control. On a first level, a space A of articulatory parameters controls continuous acoustic spectra belonging to a spectral space $\mathcal{S}$. One has thus a first field $\alpha : A \rightarrow \mathcal{S}$, directly observable and modeling the spectral variations relative to articulatory variations (see the classic works of P. Delattre and the Haskins Laboratories). Yet on a second level, the auditory transforms of the acoustic spectra constitute a space $\mathcal{F}$ which controls a space $\mathcal{P}$ of mental processes defining the percepts. Thus we have a second field $\sigma : \mathcal{F} \rightarrow \mathcal{P}$ not directly observable and whose space of control is not a space of parameters but itself a functional space of *forms*. The two levels are connected on the one hand by the auditory transforms $T : \mathcal{S} \rightarrow \mathcal{F}$ and on the other hand by a feed-back $\varphi : \mathcal{P} \rightarrow A$ indicating that perception reconstructs also an articulatory motor-schema and that, correspondingly, articulation is controlled by a space of internalized phonemic targets. In this general model,

$$A \xrightarrow{\alpha} \mathcal{S} \xrightarrow{T} \mathcal{F} \xrightarrow{\sigma} \mathcal{P}$$
$$\xleftarrow[\varphi]{\qquad\qquad\qquad\qquad}$$

the distinctive features can be conceived in *four* different ways:

(i) as frames adapted to the boundaries induced into A either by $\mathcal{S}$, $\mathcal{F}$ or $\mathcal{P}$, i.e., as external features of articulatory nature, which would be the trace of instabilities and conflicts of invariants of acoustic spectra, of their auditory transforms or of their perceptual images;

(ii) as properties of qualitative types of acoustic spectra, i.e., as internal features of acoustic-auditory nature, characterizing elements of $\mathcal{S}$ or $\mathcal{F}$;

(iii) as frames adapted to the boundaries induced into $\mathcal{S}$ or $\mathcal{F}$ by the perception $\mathcal{P}$, i.e., as external features of auditory-acoustic nature which would be the trace of instabilities and conflicts of 'attractors' of neural dynamics defined on spaces of very high dimension; and finally,

(iv) as invariants of these 'attractors', i.e., as internal features of a perceptual nature.

b. The discretization condition

The *a priori* of categorization enables us to overcome the antinomy between the discrete and continuous dimensions, and to unify the two conceptions of phonemes as

(i) prototypes for classes of tokens (allophones): logical criterion of identity,

(ii) domains in a control space: positional criterion of identity.

Using a 'geographical' metaphor, we can say that phonemes as prototypes are just like the 'capitals' of the domains defined by the categorization K of W. Moreover, the *a priori* reveals an irreducible conflict between the logical and positional criteria of identity. For these criteria to be compatible, a condition which we call the *discretization condition* has to be satisfied. It says that there exists a one-to-one correspondence between the prototypes and the connected components of the complement $W–K$ of K in W. It is fulfilled by the categorization of fig. 6 (see fig. 7).

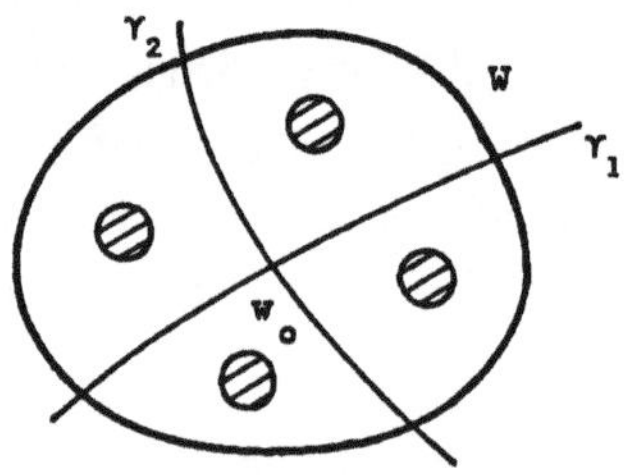

Fig. 7. The discretization condition. There exists a one-to-one correspondence between the 'capitals' and the connected components of $W–K$ ($K = \gamma_1 \cup \gamma_2$).

Yet, there is no reason at all why it has to be necessarily satisfied. For example, let us consider in a 2-dimensional control-space W, a boundary K with an end point δ. K is a threshold separating two determinations A and B by way of a qualitative opposition A/B. Now, at δ, the threshold disappears and a neutral-complex term $A{*}B$ is generated. Hence, there exist *three* prototypes ($A, B, A{*}B$) for a *single* connected component. This type of boundaries cannot be eliminated from the theory. In the case of phase-transitions in physics (which, as we have seen, is a model for the explanation of categorical perception), it corresponds to the existence of *critical points* (for instance the end point of the liquid/gas interface for water). In phonetics, it corresponds to descriptions in terms of distinctive features of the type:

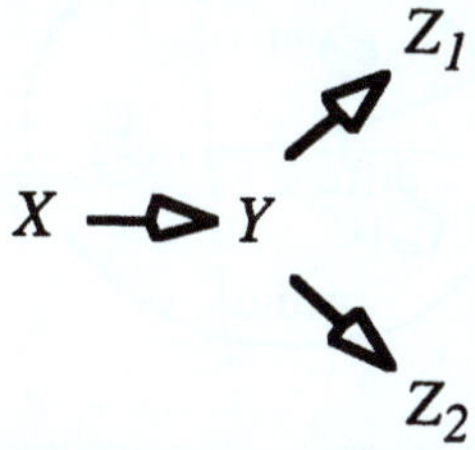

that is, to the subordination of an opposition Z_1/Z_2 to one of the terms Y of a dominant opposition X/Y. *This is in particular the case for the well-known Jakobsonian description*:

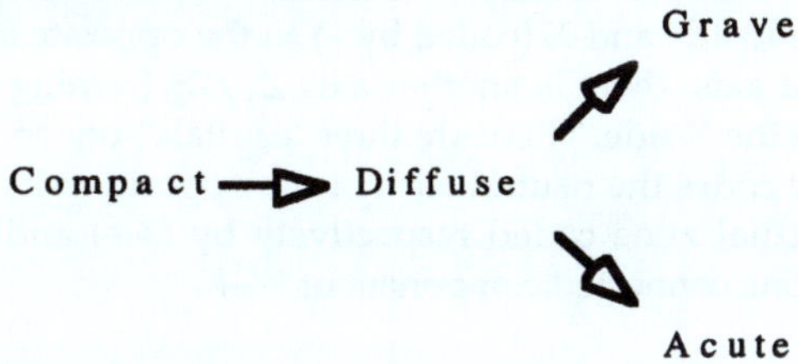

of the universal vocalic triangle (see fig. 8).

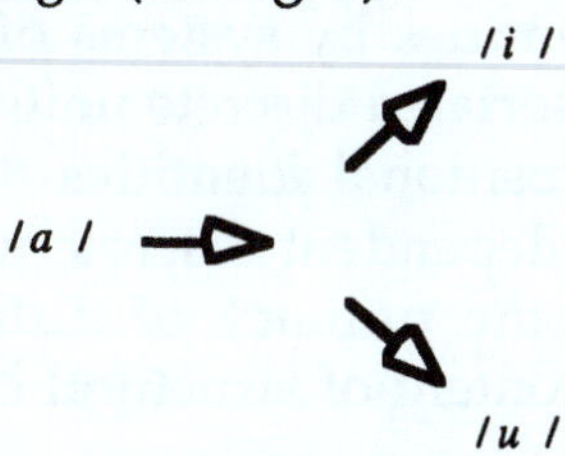

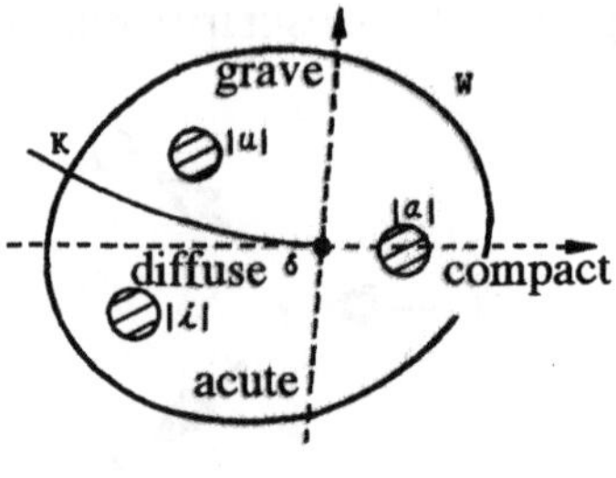

Fig. 8. A case where the discretization condition is not fulfilled. The boundary K stops at the 'critical point' δ (vanishing of the threshold separating the two domains). The 'adapted frame' is constituted at first by a dominant axis tangent to K at δ, axis divided into two zones: Y (coded for example by +) corresponds to the K side and X (coded by –) to the opposite side. Subordinated to this dominant axis, there is another axis Z_1/Z_2 framing the two zones separated by K on the Y side. There are three 'capitals', one in the zone X coded by (–,0) (where 0 codes the neutral term of the opposition Z_1/Z_2), the two others in the conflictual zone coded respectively by (+,–) and by (+,+). However, there is only one connected component of W–K.

This is a general (and fundamental) phenomenon. The determination of positional identities by systems of oppositions may come into conflict with the associated discrete units and violate the classical principle of identity. Positional identities do *not* necessarily correspond to *individuals* (independent discrete units). This validates the structural principle of the primacy of difference over identity. It shows that the eidetic content of structural categories cannot be of a naïve logical nature.

c. *The mathematics of phenomenological abduction*

The catastrophist approach to categorization *synthesizes* the two main conceptions of phonetics, namely, the *substance-based* and the *form-based* ones. The key idea is that the form (of expression) emerge from the organization of substance. To get a unified *phonetico-phonological* theory, we have:

(i) to conceptualize the relational form of expression as a logico-combinatorial description of the discriminant morphologies (W, K), and

(ii) to assume a dynamical mechanism X (neurophysiologically implemented) generating these morphologies.

It is here that CT becomes, as we have shown in Sec. I.5, a fundamental tool. It provides an advanced *mathematical* analysis of situations of control $\sigma : W \to \mathcal{X}$, and show that:

(i) the mechanisms X generating the morphologies (W, K) are highly *underdetermined* by the geometry of K;

(ii) therefore, the morphologies are to a large extent *independent* of the substrata they are implemented in, and subjected to purely geometrical constraints;

(iii) it is possible to deduce from the geometry of K constraints on X, and, in the simplest cases, it is even possible to deduce from K a *minimal* generating mechanism X_0 which the real generating mechanism X complexifies.

In Sec. I.5.2a, we have called the latter principle, the *principle of phenomenological abduction*. It reverses the order of deduction usual in science. Instead of causally deducing K from an explicit (e.g. neurophysiological) knowledge of X, one attempts, on the contrary, to abduce *from* K a partial knowledge of the unknown X. The two essential features of catastrophe theory, namely independence from the substratum and phenomenological abduction, allow us to understand how the relational form of expression can be ontologically autonomous even if it emerges from the substratum matter. This is what we call a *morphodynamical fonctionalism*.

1.2.7 *Precatastrophic interpretations of categorical perception*

In an attempt to discover the eidetic content of categorical perception, several authors had arrived at conclusions quite similar to that of CT.

a. *Prototype models*

Given a phonetic categorization (W, K), allophones in the neighborhood of K are, in general, not naturally producible, and have to be synthesized. The naturally articulated allophones cluster in the central zones (like 'capitals') corresponding to *typical* values of acoustic cues.

Indeed, as pointed out by B. Lindblom, the articulatory process is output-oriented (i.e., driven by perceptual targets). These targets are chosen in order to maximize the acoustic stability relative to the articulatory variability. This is one of the main features of what Lindblom calls the 'distinctiveness condition'.[1] Similarly, Peter Ladefoged suggested 'interpretation conventions' associating the ideal form of discrete linguistic units with typical values of control parameters.[2]

According to this point of view – which D. Massaro called 'template matching scheme'[3] – phonemes are conceived of as 'Gestalten'. It is assumed that there exists in the space P of percepts (endowed with a perceptual distance), a finite number of phonetic patterns, or prototypes, $p_1, \ldots, p_n$. Given a percept p, a variational algorithm optimizes the distance from p to the p_i and recognizes (identifies) p as a token (an allophone) of the prototype p_i whose distance to p is minimal. The categorization K thus consists of a set of median hyperplanes (segments in dimension 2). We call this a T-classification (as opposed to the catastrophist K-classifications) (see fig. 9).

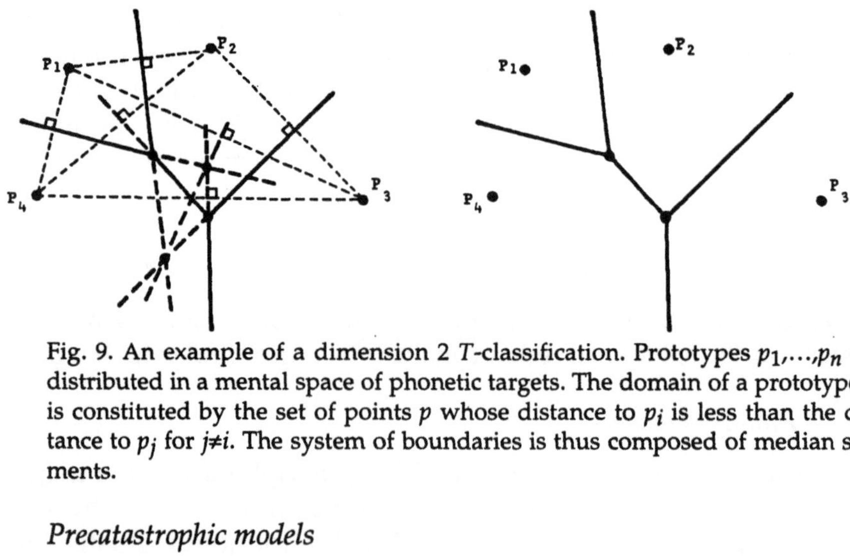

Fig. 9. An example of a dimension 2 T-classification. Prototypes $p_1, \ldots, p_n$ are distributed in a mental space of phonetic targets. The domain of a prototype p_i is constituted by the set of points p whose distance to p_i is less than the distance to p_j for $j \neq i$. The system of boundaries is thus composed of median segments.

b. Precatastrophic models

Along with the prototype models associated with T-classifications, models of a rather pre-catastrophic nature were developed in asso-

---

1 Lindblom, 1972.
2 Ladefoged, 1972.
3 Massaro, 1972.

ciation with *K*-classifications. They are based on the following hypotheses:

(i) the continuous acoustic spectra are controlled and deformable forms;

(ii) in the auditory processing, a *qualitative* information is retrieved from these spectra;

(iii) some types of spectra are *stable* relative to their control, and others are not;

(iv) perception selects invariants from qualitative spectral types;

(v) categorical perception results from catastrophic transitions of these qualitative types under the action of control. As noted by D. Pisoni, there can be categorical perception wherever complex entities controlled by parameters contain several distinctive qualities whose presence or absence defines different domains of control.[1]

Thus the basic idea is to treat the percepts

> not as bundles of separately extracted phonetic features but as integral dimensional entities whose dimensions are inseparable aspects of the whole pattern. [...] The dimensions are assumed to reflect the auditory properties of the stimulus and thus are continuous, not binary. Instead of representing speech sounds as matrices of discrete feature values, they are conceptualized as points in a continuous multidimensional perceptual space.[2]

Thus, we no longer postulate that perception depends on a detection of acoustic cues which subsequently combine into a unified percept, but that it is based on a control and that

> the sound simply initiates the unfolding of a complex pattern of neural response that directly supports the phenomenal experience.[3]

This structural (holistic) point of view makes it possible to understand categorical perception without subordinating sensorial discrimination to a cognitive identification. Indeed, only if we adopt a

1 Pisoni, 1979.

2 Repp *et al.*, 1978.

3 Bryant, 1978: 616.

prototype model, the identification is cognitive, since it results from a matching with patterns stored in a long term memory. Instead, if we assume that identification is based on the value of qualitative invariants, then it becomes a low level categorical discrimination. Moreover, in order to explain in this perspective the existence in preverbal children of innate capacities of discrimination, categorization, and perceptual invariance, it is sufficient to assume that there exists, in the initial state of the perceptual system, genetically constrained routines involving an innate field $\sigma_0 : \mathcal{F} \to \mathcal{P}$.

Such an hypothesis is in agreement with Eimas' assertion that

> the fact that constancy is present in prelinguistic infants is strong evidence that the means by which it is achieved is a result of biological constraints on the infant's perceptual system, such that the system must either be innately atuned to the invariance in the signal or innately able to impose constancy on acoustic diversity.[1]

Yet, this does not imply at all (as Eimas seems to believe) the existence of feature detectors. We would rather agree with Jusczyk that the initial field $\sigma_0 : \mathcal{F} \to \mathcal{P}$ – which is a psychophysical universal of the human species, phylogenetically inherited from the auditory mechanism of higher mammals – determines the properties of perceptual *saliency* for the acoustic stimuli to which our language clings.

> As the child begins to acquire the phonological structure of the language, one would expect to see him weight the various acoustic cues present in the speech signal according to their salience in marking distinctive contrasts in the language.[2]

During language acquisition the *universal* K-classification K_0 determined by σ_0 will be deformed, complexified and specified by the linguistic environment and progressively subordinated to a *T*-classification of a cognitive nature.

However, it is certainly Kenneth Stevens who came closest to the catastrophist formulation of categorical perception, especially in his classic paper *'The Quantal Nature of Speech'*.[3] Stevens studied the

---

1 Eimas, 1980.
2 Jusczyk, 1980.
3 Stevens, 1972a.

first level of control defined by the field $A \xrightarrow{\alpha} \mathcal{S} \xrightarrow{T} \mathcal{F}$ and introduced two basic ideas:

(i) The control map between A and $\mathcal{S}$ (i.e., between the articulatory and the (audio)-acoustic levels) by the control is 'non-linear'. There are domains of A, whose associated spectra are stable with respect to the control α and constitute the basis of the phonological code.

> For a particular range of an articulatory parameter, the acoustic output from the vocal tract seems to have a distinctive attribute that is significantly different from the acoustic attributes for some other region of the articulatory parameter. Within this range of articulation, the acoustic attribute is relatively insensitive to perturbations in the position of the relevant articulatory structure.[1]

(ii) The acoustic attributes are *spectral configurations of higher order*, i.e., *qualitative and global* properties of the spectra, and not isolated elementary cues. In other words, they are properties of the 'gross shape' of the spectra or, more precisely, of their auditory transforms.[2]

Hence, according to Stevens, the relation between an articulatory parameter and the acoustic attribute which it controls, is typically catastrophist, as is shown in fig. 10, taken from Stevens, 1972.

---

1 See MacNeilage, 1979.
2 Stevens, Blumstein, 1978.

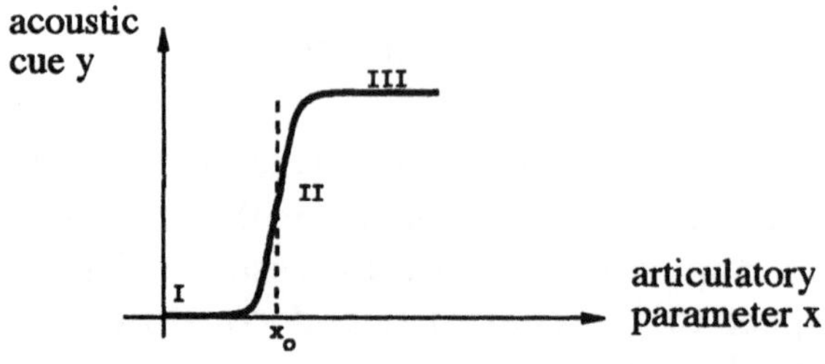

Fig. 10. A typical relation between an articulatory parameter x and an acoustic cue y. In the domain I ($x<x_0-\varepsilon$), the cue y is absent. In the domain III ($x>x_0+\varepsilon$), y is present and insensitive to variations of x. In the domain II ($x_0-\varepsilon<x<x_0+\varepsilon$), y is unstable relative to the variations of x and a 'catastrophic' transition happens.

Stevens is very explicit:

> There are certain conditions for which a small change in some parameter describing the articulation gives rise to an apparently large change in the acoustic characteristics of the output, there are other conditions for which substantial perturbations of certain aspects of the articulation produce negligible changes in the characteristics of the acoustic signal.[1]

This explains the acoustic-articulatory basis of the phonological form, avoiding at the same time a physicalist elimination as well as a structural realism of forms (see Sec. I.2.6c).

1.3 *Principles of catastrophist modeling*

Catastrophe theory is concerned with controls maps $\sigma : W \rightarrow \mathcal{X}$ wherever the internal processes are defined by *dynamical systems* (vector fields) X_w on differential manifolds M. As this research program is too complex, since the structure of generic dynamical systems is still widely unknown, in the 'elementary' theory it is restricted to dynamical systems X_w *derived from potential functions* $f_w : M \rightarrow \mathbb{R}$ ($\mathbb{R}$ being the field of real numbers). For a given n-dimensional manifold M (assumed to be compact) we have to explain the structure of:

(i) the space $\mathcal{F}$ of differentiable maps $f : M \rightarrow \mathbb{R}$, and

1 Stevens, 1972a: 52.

(ii) the fields $\sigma : W \to \mathcal{F}$ mapping the control space W on $\mathcal{F}$.

The main issue is to understand the *geometry* of the catastrophic sets $K_{\mathcal{F}}$ and K. We present briefly below some of the key ideas.

1.3.1 *The qualitative type*

The qualitative type of a potential $f : M \to \mathbb{R}$ is expressed by its differential type, defined by the action of the group $G = G_M \times G_{\mathbb{R}}$, where G_M (resp. $G_{\mathbb{R}}$) is the group of diffeomorphisms (i.e., of automorphisms for the differentiable structure) of M (resp. of $\mathbb{R}$). If $f \in \mathcal{F}$, its orbit f^* under the action of G is thus made up of the functions $g = \psi \circ f \circ \varphi^{-1}$, where $\varphi \in G_M$ and $\psi \in G_{\mathbb{R}}$.

If $f \in \mathcal{F}$, the essential qualitative (i.e., G-invariant) information on its structure is provided by its *critical elements*. Let $x \in M$ and let $Df(x)$ be the linear tangent map of f at x. Df maps linearly the tangent vector space T_xM of M at x on the tangent vector space $T_{f(x)}\mathbb{R}$ of $\mathbb{R}$ at $f(x)$. If $(x_1, \ldots, x_n)$ is a system of local coordinates at x, the matrix of $Df(x)$ is the $1 \times n$-matrix $(\partial f/\partial x_1, \ldots, \partial f/\partial x_n)$. We say that x is a *critical point* of f if $Df(x)$ is not of maximal rank (namely 1) at x, i.e., if all the partial derivatives $\partial f/\partial x_i$ of f vanish at x. $y = f(x)$ is then called a *critical value* of f. If y is a non-critical (i.e., regular) value, then f is 'locally trivial', which means:

(i) the pre-image $f^{-1}(y)$ of y by f is a sub-manifold M_y of M of codimension 1 (i.e., of dimension $n-1$), and

(ii) there exists a neighborhood U of y, such that $M_U = f^{-1}(U)$ is diffeomorphic to the direct product $M_y \times U$ (see fig. 11).

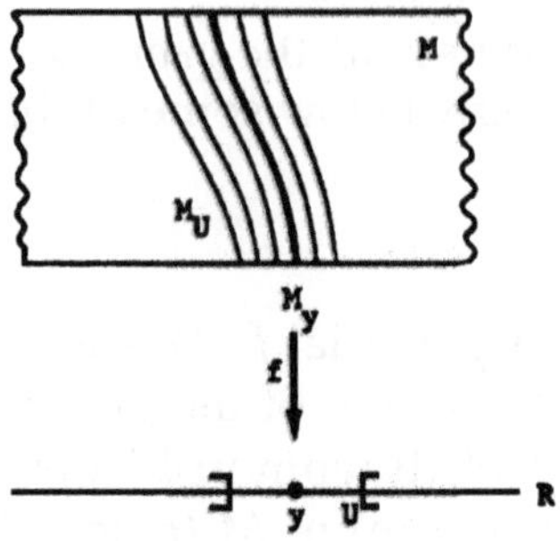

Fig. 11. Local triviality of a potential function $f: M \to \mathbb{R}$.

When x is a critical point, its 'degree' of criticality is measured by the degree of degeneracy of the higher partial derivatives of f at x. The second derivative of f is given by the symmetric $n \times n$ matrix of the second partial derivatives, called the Hessian:

$$H = \left(\frac{\partial^2 f}{\partial x_i \, \partial x_j} \right).$$

If H is of maximum rank (namely n), then we say that x is a *non-degenerate* critical point, and we define its *index* as the index of the quadratic form H. The critical elements of f are its critical points with their 'degree' of criticality, and its critical values with their multiplicity.

1.3.2 *The criterion of structural stability*

A potential $f \in \mathcal{F}$ is structurally stable if every potential g sufficiently close to f (for the differentiable topology) is G-equivalent to f or, in other words, if the orbit f^* of f contains a full neighborhood of f. A fundamental theorem, namely *Morse theorem*, asserts that, if M is compact, f is structurally stable if and only if:

(i) all of its critical points are non-degenerate; and

(ii) all of its critical values are distinct.

Thus in that simple case there are only *two* possible causes for instability: the degeneracy of critical points, and the equality of critical values. They correspond respectively to the two main types of catastrophes: the bifurcation catastrophes, and the conflict catastrophes.

1.3.3 *Genericity and transversality*

The equivalent of the Taylor series of a function $f: \mathbb{R} \to \mathbb{R}$ is given by the maps $j^k f: M \to J^k$ – called jet maps – which associate to each $x \in M$ the sequence $j^k f(x)$ of the partial derivatives till the order k of f at x (these sequences are called the k-jets of the map and generate fibered spaces J^k with base space $M \times \mathbb{R}$). Now, it is easy to verify that the first condition of Morse theorem says exactly that the 1-jet $j^1 f(x)$ of f at x is *transversal* to the zero section of J^1 ($j^1 f(x) = 0$ says that x is critical and transversality expresses the non-degeneracy). The second condition can likewise be expressed in terms of transversality. By virtue of a fundamental theorem due to René Thom, this implies that structural stability is a *generic* property of potentials, i.e., that non-stability is 'exceptional'.

1.3.4 *Finite determination, universal unfoldings and transverse models*

Let $K_{\mathcal{F}}$ be the global catastrophic set of $\mathcal{F}$. It is the set of structurally unstable elements. If $f \in K_{\mathcal{F}}$, the geometry of $K_{\mathcal{F}}$ in a neighborhood of f *classifies* the stable types that can be obtained from f by small deformations. If f is 'infinitely' unstable (for instance, if f is constant), the local geometry of $K_{\mathcal{F}}$ at f will be 'chaotic'. The elementary theory is interested in the 'good' cases of 'weak' unstability where the following conditions obtain:

(i) f is *finitely determined*, i.e., f is equivalent to one of its jets of finite order (which is a polynomial). This implies that f is of *finite codimension*, i.e., the orbit f^* of f admits at f a finite dimensional complementary space W in $\mathcal{F}$.

(ii) In a neighborhood of f, the pair $(\mathcal{F}, K_{\mathcal{F}})$ is equivalent to the direct product of the pair (W, K) (where $K = K_{\mathcal{F}} \cap W$), with the orbit f^* (see fig. 12).

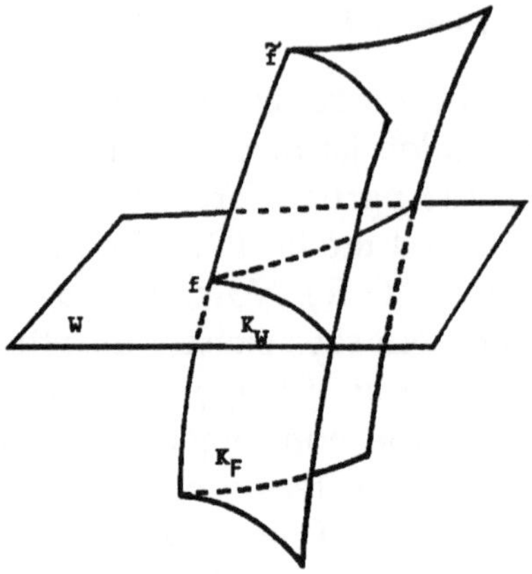

Fig. 12. The locally trivial structure of $K\mathcal{F}$ in a neighborhood of a 'good' finitely determined singularity.

In this case we say that (W, K) is a *transverse model* of f. As W is of finite dimension k, it is isomorphic to a neighborhood W' of the origin of $\mathbb{R}^k$ and thus we can interpret W as a field – an *unfolding* – $\sigma : W' \to \mathcal{F}$ which associates to $w \in W'$ the corresponding element $f_w \in W$. σ is called a *universal unfolding* of f, and f is called its *organizing centre*. The universal unfolding theorem essentially states that σ is *unique* up to equivalence and allows to reconstruct *all* the unfoldings of f.

1.3.5 *Degrees of instability and stratifications*

If $f \in K\mathcal{F}$ corresponds to the previous elementary situation, the possibility of progressively stabilizing it by successive steps is readable on the *geometry* of K for any universal unfolding (W, K): K will be *stratified*, that is to say, composed of a piling up of singular loci of decreasing dimension, each stratum corresponding to a precise degree of instability. These stratifications *geometrize the concept of classification.*

1.3.6 *Normal forms and the theorem of classification*

Thom's theorem on elementary catastrophes is a theorem of classification (up to differentiable equivalence) of the singularities of codimension ≤ 4. For every case, it yields:

(i) the number of internal variables (1 or 2) effectively involved in the instability of f;

(ii) a normal (polynomial) form of the generating singularity (organizing center), i.e., the simplest representative of its equivalence-class;

(iii) a normal form of its universal unfolding.

1.3.7 *Methodological rules for modeling*

Given a phenomenon manifested by a morphology K in a control space W, the methodology of CT consists, as we have seen, in *abducing* from K constraints on the generating internal mechanism X. It is based on certain principles, of which we give two examples:

(i) The morphology of elementary catastrophes C being known, if one of these C is empirically oberved, then one can assume that its internal dynamics X_C governs the phenomenon, and that the real generating mechanism X is a complexification of X_C.

(ii) On the other hand, given a field $\sigma : W \rightarrow \mathcal{X}$, it must be structurally stable for the associated phenomenon to exist. This stability is expressed by the *transversality* of σ to $K_{\mathcal{X}}$. Now, a consequence of this *a priori* transversality is that $\sigma(W)$ must *avoid* the strata of $K_{\mathcal{X}}$ whose codimension is greater than the dimension of W. *The dimension of a control space W drastically limits the complexity of the morphologies which can be unfolded in it in a structurally stable way.*

1.4 *From the second to the first articulation of language*

In conclusion, we see that, as mathematical models of critical phenomena, catastrophist models are also *morphodynamical models for paradigmatic categorization.* In this sense they are particularly well-suited for structuralist theories. Having sketched their relevance for categorical perception in phonetics (second articulation of language), we will now proceed to the other 'pillar' of structuralism, namely the actantial conceptions of syntax (first articulation).

2 Actantial schematism

We shall pursue in this section, an enquiry similar to the one we have just undertaken for phonology. Instead of the distinctive features belonging to the second articulation of language, we will be concerned with the actantial relations belonging to the first articulation. We will see that this latter notion is also based on reciprocal presupposition of positional values, and that it also needs a geometrical schematization. However, the situation is here radically different since, contrary to the case of phonetics, the schematization is no longer correlated with modeling physical data. We will address this difficult problem in the following way.

2.1 *Towards a pure 'etic' linguistics*

We have already emphasized, in our presentation of the 'foundational aporia' of phonetics, the interplay of effective modeling and foundational issues. Contrary to general belief, though the social sciences succeeded, each in its own way, in *delimiting* their objects of study, they have not been able to *constitute* them as such. They combine in general four theoretical approaches:

(i) conceptual systematization of empirical corpus;

(ii) general methods of data analysis (e.g., factorial analysis, time series, etc.);

(iii) abstract formal modeling (e.g., AI models);

(iv) reflexive interpretations (e.g., hermeneutics).

If we take mathematical schematization to be an *objective determination* in a strong sense, then we see that *none* of the above theoretical approaches constitutes an objective determination: the first, because it does not go beyond the empirical and the conceptual levels; the second, because it does not capture the internal structure of the object; the third, because it relies on purely general symbolic forms; the fourth, because it uses interpretative methods opposed to the very idea of objectivity.

In this context, the catastrophist schematization has a special methodological relevance. Let us further clarify this point. The object of phonetics, as we have seen, is both psychophysical and linguistic. When CT treats morphodynamically categorical perception as a perceptual case of critical phenomena, it operates as a *modeling* device for the psychophysical *phonetic* domain. But when it treats categorical perception at the purely structural level, it operates as a *schematization* for the *phonological* domain. In other words, if CT has the capacity to overcome the 'foundational aporia' of phonetics, it is because it can operate as a *mediation* between the psychophysical phonetic objectivity, and the structural phonological objectivity. This explains why the applications of CT are twofold. While, as modeling, its concepts have an empirical value and refer to an already existing objectivity, as schematization, they operate as a constitutive stance for a new type of objectivity.

Now, let us imagine that we were schematizing a structural phonology devoid of phonetics. We would then have developed a similar thesis on the relevance of a catastrophist schema, but its justification would have been very different. We would no longer have been able to develop the mediation between phonetic modeling and phonological schematization, and to further legitimize – except by referring to problematic *a priori* evidences – the interpretation of distinctive features as critical phenomena, or that of phonological hierarchies and markedness as stratifications.

Now, it is precisely such a difficulty that we run into while tackling the first articulation of language. *Structural syntax has no observable correlates in the manner that structural phonology has phonetic psychophysical correlates.* That is why the algebraic-combinatorial symbolic nature of syntax is given primacy.

In this section we will question this formalist assumption, which is, in fact, unjustified. Just like the phonetic one, the syntactic object is 'bimodal' and thus must be analyzed at two different levels of objectivity. It is bimodal because there are two well-defined types of syntactic relations: the grammatical relations, and the 'semantic' case relations , i.e., the *actantial* relations defining the *semantic roles*. Now, these two types of relations come under quite different types of theoretical, methodological and epistemological concerns. Grammatical relations belong to language automatisms. They involve the algorithmic aspect of syntax (as modeled for instance by generative grammars). Actantial relations, on the other hand, bear the semantic roles situated at the interface of language and thought. They are not concerned with the open-ended recursive rules of syntax, but on the contrary *with closed classes of conceptual relations*.

The bimodality of the syntactic object can be better understood as *a complementarity between grammatical generativity and actantial closure*. To describe such a complementarity, we have to work out *a double theory*. A structural syntax reduced to a generative theory and devoid of any actantial theory is only a *half*-syntax. That is why we propose to develop in this section an examination on the *eidetic* content of the actantial conception. We will try to show, that the inherent complementarity of the syntactic object, far from being acknowledged, is instead systematically ignored. We will see that actantial relations are relational morphologies pertaining to a geometry of position and we will suggest a schematization with constitutive power.

There are two quite different aspects of language. The first involves 'the adequate description of linguistic activity as understood cross-linguistically'.[1] It is a matter of systematic empirical linguistics. The second involves not as much the detailed organization of this activity, but its conditions of possibility, its ontogenesis, its rooting in perception and action, and also the constraints that the structure of the objective world imposes on it. It is a matter of 'pure' linguistics conditioning systematic empirical linguistics. And without constituting the former, the latter cannot attain a properly *theoretical* status. Indeed, the systematic organization of an empirical diversity is not sufficient to constitute the regional ontology of a theoretical object.

Chomsky's rationalist linguistics represents the first successful attempt at a 'pure' linguistics. It is primarily concerned with the concept of a grammatical rule; the eidetic character of language it tries to formalize is that of generativity; and it takes mathematics (automata theory, formal grammars, etc.) as constitutive of empirical linguistic phenomena. However, as it has become increasingly evident,[2] Chomskyan perspective can cope with systematic empirical linguistics only by considerably weakening its basic theoretical

---

1 Culioli, 1970: 2.

2 See below, Sec. 2.2, and 4.1.

underpinnings. This inherent limitation results from not seriously considering certain phenomenological features of natural languages.

We can describe this inherent limitation of the formalist conceptions of language as follows. If we take a grammar as a set of generative mechanisms for associating semantic representations with phonetic sequences, then we can regard syntax as a system of formal constraints conditioning this mediation between phonetics and semantics. In this sense, a formal grammar must assign a structural description to each sentence, and moreover generate an infinite number of such descriptions. For this purpose, it is hardly necessary to assume (as Chomsky does), in the pretransformational base component, a categorial component generating the phrase-markers of kernel sentences whose semantic interpretation determines the deep structures. In fact, the main role of the kernel phrase-markers is to provide a *configurational* definition – i.e., a positional definition in terms of dominance – of grammatical relations and functions. For this it would suffice to use tree-diagrams. To derive these by means of iterated rewriting rules is superfluous. Along with Thom, we reject the belief that the generative status of formal structures should be accepted a priori and does not require any explanation. We accept the claim that in natural languages 'it is the self-limitation of the generative capacities of syntax that calls for an explanation'.[1]

In model-theoretic logic, there exists a separation between syntax (deductibility) and semantics (denotation, truth, validity).[2] It does not work the same way in linguistics. As A. Culioli pointed out 'nothing allows to reduce the semantics of natural languages to the interpretative semantics of formal systems'.[3] We can even think

> that there exists at a very deep level (presumably prelexical) a grammar of primitive relations where the distinction between syntax and semantics makes no sense.[4]

---

1 Thom, 1971.

2 For an introduction to the logical theory of models (theorems of Löwenheim-Skolem and of Gödel, ultra-filters and ultra-products, non-standard arithmetic and analysis, etc.), see Petitot, 1979a.

3 Culioli, 1970: 7.

4 Ibid.: 8.

But if this is true, then syntax of natural languages is of an eidetic type different from that of formal syntax, and therefore mathematical linguistics should transform its conception of 'pure' linguistics, and rethink its relationship with systematic empirical linguistics. It has to complement formal 'pure' linguistics with another 'pure' linguistics no longer concerned with the algebraization of the competence rules, but with *the schematization of primitive syntactico-semantic relations in the framework of a dynamical theory of performance.*

In two seminal papers,[1] René Thom proposed to order the grammatical categories on a bidimensional squish.[2] On the *X*-axis of the squish the categories are ordered as follows:

Nouns-Verbs-Adjectives-Numerals-Possessives-Deictics-
Logical functors-Quantifiers.

If we now put on the *Y*-axis the semantic variability of the categories (i.e., the range between the maximal concretion and the maximal abstraction of their occurrences), we notice the following facts.

(i) The semantic variability decreases along the squish and collapses when crossing the numeral zone and shifting from open classes to closed ones. It is very high for nouns and verbs, and nil for the logical functors.

(ii) The squish extends from the 'categorematic' pole to the 'syncategorematic' one. Following Pike's etic/emic opposition, Thom hypothesized that the first pole is an 'etic' objective one linked to the simulation of external reality by language while the second is an 'emic' subjective one linked to the automatisms of competence:

> We see that linguistic entities [...] are of very different kinds. With the noun, we are dealing with an entity endowed with a certain autonomy: the referent occupies a portion of space, which it defends against environmental perturbations [...]; on the contrary, the grammatical auxiliaries, owe their meaning only to a quasi-ritualized activity of the speaker; they are totally embedded in the automatisms of language.[3]

1 Thom, 1973b, 1978b.
2 The term 'squish' was introduced by John Ross.
3 Thom, 1978b: 79.

(iii) The central zone of the squish where the semantic variability breaks down represents a sort of *threshold* between the etic-objective and the emic-subjective poles.

These remarks enable us to understand why contemporary formal linguistics is only a half-linguistics. It is based on an *emic* 'pure' linguistics devoid of any 'etic' dimension. Thus, we are left with the necessity of developing mathematically the principles of *a 'pure' etic linguistics,* establishing their complementarity with pure emic linguistics, and linking them up with systematic empirical linguistics.[1]

In 'pure' etic linguistics, primacy is accorded to the *regulation* of the three basic grammatical categories,[2] namely, nouns, verbs, and adjectives.

(i) As regards the regulation of nouns, one assumes that

> there exists a certain isomorphism between the psychological mechanisms ensuring the stability of a concept *Q*, and the physical mechanisms ensuring the stability of the real object *K* represented by *Q*.[3]

Therefore, the referential function is conceived of not as a correspondence between language and reality but as a constraint imposed on semantics by the reality it simulates. A concrete concept is a dynamical psychological entity, whose regulation figure (*logos* in Thom's sense) is partly *isologous* (isomorphic in the sense of logoï) with that of the referent. Such an assertion amounts to a phylogenetic hypothesis on language: the earliest concepts must have been of those entities whose recognition was fundamental for survival (prey, predator, sexual partner), and that is why, according to Thom, 'the logos of living beings served as a universal pattern for the formation of 'concepts''.[4]

---

1 We will return on this fundamental etic/emic distinction in Sec. III.2 to discuss some theses proposed by Thomas Pavel.

2 These categories (parts of speech) are considered here not as entities to be constructed, but as prelinguistic universals imposed by the phenomenology of reality.

3 Thom, 1973b: 247.

4 Thom, 1980a: 131.

(ii) The regulation figure of a concept C is intimately linked with its verbal spectrum. The regulation catastrophes which delimit the domain of existence of C in its substratum space are identifiable with the verbal interactions in which C can participate as an actant. We encounter here a neo-Tesnierian conception of the verb as an 'organizing centre', i.e., as an event distributing actantial places. As we will see later, the basic function of verbs is to simulate the elementary actantial interactions realizable in space-time.

(iii) Finally, the regulation of adjectives which connect substantives to qualitative spaces is reduced to the *categorization* of these spaces, e.g., the color domain.

The development of 'pure' etic linguistics thus needs a mathematical theory of

(i) conceptual regulation,

(ii) verbal valence, and

(iii) categorization.

It must, moreover, be correlated with the cognitive organization of perception and action. Indeed, as Charles Osgood and Alexander Luria have observed:

> it seems perfectly reasonable to think that much, if not all, that is universal in human language is attributable to underlying cognitive structures and processes. [...] Perceptual and linguistic signs and sequences must, at some level, share a common representational (semantic) system and a common set of organizational (syntactic) rules, cognitive in nature.[1]

> We must look for the roots of basic linguistic structures in the relations between the active subject and reality and not in the mind itself.[2]

As Wolfgang Wildgen stressed, this requirement is at the core of Catastrophe Theoretic Linguistics:

> We assume that the dynamic principles governing the semantics of words are intrinsically connected with basic propositional structures. This is especially true for verbs. Our dynamic [i.e., catastrophist] treatment of verbs starts with

1 Osgood, 1971.
2 Luria, 1975.

> a consideration of the dynamic principles underlying the perception of space and time and of changes, motion, locomotion, and action in space and time. [...] In a general semantic theory our archetypal and dynamic component would be a basic stratum whose influence becomes weaker as we progress to the levels of syntax and text (conversation).[1]

So, from an etic perspective, *the linguistic object is no longer autonomous*. The regulation of nouns is linked to biological regulation, the verbal valence to a physics of actantial interactions, and the categorization of semantic spaces to critical phenomena similar to phase-transitions. This loss of autonomy of the linguistic object may appear unacceptable to many linguists. However, it is necessitated by empirical experience. It reorients linguistics in a more naturalistic and cognitive direction. We will begin confirming it with a critique of the generativist paradigm.

2.2 *Five limitations of the Chomskyan paradigm*

The Chomskyan theory of language privileges the rules of syntactic competence and the algorithmic dimension common to natural and formal languages.[#] In our view, it suffers from certain *intrinsic* limitations, of which we will discuss five.

1. Chomskyan theory transfers from formal to natural languages the idea that generativity is free and can operate without restriction. Now, the basic concepts linking syntax and semantics in model theoretic logic make sense only if we can consider an *infinite* set of sentences. Metaphorically, we can say that they are adapted for the description of the 'asymptotic states' of formal languages. But natural languages have no 'asymptotic states'. Their generativity is not free. It is drastically *constrained*, and its 'recursivity' is of very limited range. In fact, natural syntax is not at all recursive in the technical sense of the term. As Maurice Gross emphasized in his impressive work *Méthodes en syntaxe*[2] concerning the structure of verbs and comple-

---

1 Wildgen, 1981: 235.

\# In the seventies.

2 Gross, 1975.

tive constructions, the creativity of language is largely due to lexical combinatorics and transformational operations of kernel sentences.

> Chomsky [...] provided a clear hypothesis on the source of creativity. He observed that natural languages are infinite sets (recursively enumerable) and that this infinite character is due to the existence of diverse recursive syntactic mechanisms which combine sentences in a manner not bounded a priori. [...] We do not fully agree with Chomsky's interpretation and formalization of the notion of creativity. We believe that, due to the combinatorial possibilities existing even at the level of simple sentences, the notion of creativity could very well find a place within the framework of finite processes. [...] Moreover, the recursive syntactic mechanisms which increase the length of sentences do not seem to make any contribution to creativity. It is possible to describe them almost entirely in terms of concatenations of simple sentences with independent contents.[1]

Instead of starting from the primitive concept of a derivation rule, a 'good' syntactic theory must explain the dramatically limited range of iterations in natural language.

2. Further, Chomskyan theory is based on the hypothesis that there exists a *global* grammar for every natural language, while empirical data show there are only divergent *local* approximative grammars. This intrinsic limit is linked to the previous problem. The fundamental idea of generative linguistics is that a natural language is axiomatizable, and that the whole set of its sentences is derivable by the recursive application of a finite set of rules. This recursive conception masks two central problems:

(i) a short range recursivity undermines the autonomy of syntax with respect to semantics, and

(ii) if the grammar of natural language is a set of divergent local grammars, then a 'good' syntactic theory should be able to move from the local level to the global one.[2]

3. The axiomatic conception seeks to embed the grammars of natural languages in a class of formal grammars where it becomes

1 Ibid.: 18.

2 On the local/global problem, see Petitot 1979b.

impossible to characterize formally their subclass. This explains Chomsky's *innateness* hypothesis. The argument is as follows:
there are universals of natural languages

(i) with respect to which a child never makes any mistake,

(ii) which he has not acquired by learning,

(iii) which are not formally characterizable,

'therefore' they have to be genetically constrained.

In this argument, evidently the third point is the more decisive. For, the universals invoked (e.g., structure-dependent rules; bound anaphora; specified subject condition) can rather be explained by a *cognitive* conditioning of the grammatical structures. But if we reduce language to pure automatisms of competence, then we are left to interpret in innatist terms the linguistic structures which exceed the bounds of a formalist description. *It is fallacious to infer an ontological proposition from an internal limitation of descriptive formalisms. This would be valid only if the formal descriptions had been previously justified as ontologically determinant.* But it is far from being the case with generative grammars.

A structural theory of syntax must satisfy what we call the *condition of descent*: after having embedded the class of natural grammars as a subclass of the superclass of formal grammars, we should be able to descend from the superclass to the subclass, and to characterize the latter using criteria definable in the former. Generative grammars do not satisfy this condition of descent and this shortcoming is made up for by an innatist hypothesis.

4. Chomskyan theory presupposes the formal homogeneity of inputs and outputs of the transformational component. As Chomsky himself pointed out, the transformational rules must apply in sequence and therefore to entities *of the same type* as those they generate. In other words, syntax is regarded as a rule-governed complexification of initial structures (kernel sentences of the deep structure) which are *already* describable by phrase-structure trees. These initial structures cannot by definition be deduced within the transformational-generative theory and hence have to be ascribed to a separate (innate) origin. Now, there is much evidence to suggest that the

primitive structures are *not* of the same formal type as the inputs and outputs of the transformational component.

5. And finally, for the same reasons, Chomskyan theory excludes the non-grammatical syntactic relations, namely the actantial relations. These are incorporated in a lexical semantic component. Case grammars are considered as mere 'notational variants' of the standard theory. Now, this does not in anyway solve the problem of the universal semantic roles.

The five intrinsic limitations described above are closely interrelated. In order to overcome them, the 'best' theoretical strategy would consist in making the four following hypotheses:

(i) There exists a set of core archetypal Gestalt-like actantial structures determining the universal semantic roles.

(ii) There is a transformation of the *type* of syntactic structures, which enable the passage from primitive actantial relations to kernel sentences. We will call it a *grammaticalization of deep actantiality*.

(iii) The kernel sentences produced by the grammaticalization of deep actantiality can be inputs to the transformational component as conceived of by the standard theory.

(iv) The short range generativity and the divergence of local grammars discussed in Sec. 1–2 can be interpreted as the effect of a resistance of primitive actantial structures to open-ended generativity.

If we accept these hypotheses, then Chomskyan theory must be complemented by a theory *of a different eidetic type* which would be capable of:

(i) making explicit the actantial structures in terms of primitive relational morphologies, accounting for their stability and their universality, and deducing them on the basis of a principle of regulation;

(ii) modeling the grammaticalization of deep actantiality;

(iii) explaining the resistance of deep actantiality to formal generative rules;

(iv) fulfilling the condition of descent according to the above criteria.

During his famous 1975 debate with Piaget, Chomsky argued that the ignorance of the physical basis of mental structures compels us to keep to abstract characterizations but that there is nevertheless no reason to suppose that the physical structures involved are fundamentally different from other physical organs that are better understood.[1] But this abstract characterization of language does not satisfy the principle of phenomenological abduction (see, Sec. I.5.2.a). Our ignorance of the physical bases of language is no reason for not assuming them and for not adopting them as underlying non-observable causes. Indeed, we must suppose the existence of dynamical processes (neurally implemented) underlying linguistic expressions. The main difficulty is:

(i) to explicate the dynamical level;

(ii) to construct models to deduce the relational morphologies from these underlying unobservable processes;

(iii) and reciprocally, to be able to proceed from the observed relational morphologies to constraints on the underlying generating mechanisms (principle of phenomenological abduction).

2.3 *The primacy of actantial relations*

A structural theory of syntax should then unify two half-syntaxes, namely, the transformational-generative and the case-based actantial ones. For this, we need an actantial 'pure' linguistics. The actantial relations are semantic (in the Hjelmslevian sense of 'form of content'). Their content, manifested morphosyntactically by case-markers (including word-order), belongs to a cognitive semiotics of the natural world. These features explain why it is so difficult to work out case grammars. If one tries to extract universal deep cases from an empirical analysis of phrase-structures, then one cannot avoid the vicious circle of a *semantic interpretation* of grammatical relations which reduces only *paraphrastically* the surface morphosyntactic

1 TLTA, 1979: 52.

structures to deep actantiality. This difficulty can only be overcome by seeking the generative principles of deep actantiality in the *extra-linguistic reality*.

We are using here 'extra-linguistic' in a double sense. Firstly, it refers to the semiotics of the natural world and presupposes a pictorial similarity between the actantial structure of a proposition and the objective state of affairs which it denotes. Secondly, it refers to the subject's cognitive capacities, i.e., to conceptual structures. At the level of actantial relations, there exists a *common* semiosis of thought, language and reality. It is rooted in the subject's representational capacities, his perceptual activity and his sensory-motor intelligence.

Roger Brown, in his classic work on child language development,[1] came up with the following conclusions.

(i) There are five stages of syntactic construction consisting of:
- semi-sentences formed of only verbs, nouns, adjectives, deictics like pronouns referring to the self and the parents, locatives, etc.;
- sentences more complete with respect to markers of gender, number, tense and aspect, articles, determinants, auxiliaries and prepositions, etc.), that is, grammatical morphemes belonging to the 'syncategorematic' (closed classes) dimension;
- modalities operating on full sentences (e.g., interrogation, negation, imperative, passivization, etc.);
- embedded clauses;
- coordination and logical relations (and, or, but, since, if ... then, though, etc.).

(ii) The syntactic structure at stage I is basically *actantial*, even though at the expression level, it requires an early mastering of grammatical relations and topicalization rules. Experiments confirm Schlesinger's hypothesis that a child's first sentences express a *pre-linguistic cognitive* organization consisting of concepts and relations (what we called the grammaticalization of deep actantiality).

(iii) Case grammars (as proposed by Chafe and Fillmore) are therefore more suitable for the description of child's language

1 Brown, 1973.

than transformational-generative grammars or generative semantics. We can assume that children acquire a case grammar before learning a more categorial type of grammar.

(iv) The major significance of stage I seems to lie in the sensorimotor intelligence and we get an evolutionary chain: sensorimotor intelligence → cognitive level of semantic roles and actantial relations → grammaticalization.

We can thus conclude that actantial relations are quite primitive, primary in grammatical development, situated at the interface of language and thought, and originate in the active perceptual simulation of the external world.

2.4 *Actantial schematism and the localist hypothesis*

Once we accept the importance of actantial relations both at the level of pure syntax and at the level of the relationships between cognitive capacities and the morphological organization of the natural world, the whole question becomes *whether their formal content can be mathematically represented.* This is the 'burden' of all case theories. If a case theory has to avoid the vicious circle of a purely semantic interpretation of deep structures, it must reduce the *substantive* content of the universal semantic roles *to a purely configurational one.* It must become truly *syntactic,* and for this, *it must characterize formally the semantic roles from an abstract schematization of actantial relations,* just as, in the standard theory, the grammatical relations are formally characterized from dominance relations defined by *positions* in the phrase-structure trees. It must *define the semantic roles no longer formally as labeled categories but configurationally as positional relations.* The actants should be conceived of as positional syntactic values and not as substantive units.

In the linguistic tradition, it is the *localist hypothesis* that came the closest to this view. According to it, each case carries a *double* determination: a *syntactic* one involving an actant, and a *local* one involving a spatio-temporal position. To describe the actantial relations, we must first account for the ambivalence between actants and spatio-temporal positions.

As we will see in more detail in Sec. 6, the localist hypothesis is the key to the actantial conception of grammar. Indeed, *from the moment we consider spatio-temporal actants whose identity is reduced to their localization, we can equate the syntactic actantial relations with interactions between such spatio-temporal localizations*. Now, these interactions are not random. They are morphologies which allow precisely to define configurationally the semantic roles as positional values. We can *classify* them and thus deduce a finite list of case-universals. The geometric interpretation of actantial relations as interactions between spatio-temporal localizations breaks the vicious circle of the semantic interpretation of deep structures and yields a principle for schematizing deep actantiality.

It would be wrong to think that because of its reference to space-time, the localist hypothesis is naïvely introducing an externalist standpoint in linguistics. *It reduces actants to localizations just as formal logic reduces them to symbols.*

3 Tesnière's notion of verbal valence

We have seen that the real task of structural syntax is to synthesize two complementary dimensions, namely grammaticality and actantiality. This idea has been masterly developed by Lucien Tesnière in his celebrated work *Éléments de Syntaxe Structurale*,[1] from which we will present now some relevant points.

3.1 *The graphical analogy of connection*

Beyond the empirical diversity of languages, Tesnière intended to identify the principles of a pure rational syntax. According to him, syntax is the study of sentence, and a sentence is first and foremost a set of disembodied, 'incorporeal' *connections* (1.2, 1.3),[2] which do not

1 Tesnière, 1959.

2 In this section, the numbers given in brackets refer to §§ in Tesnière's *Éléments de Syntaxe Structurale*.

have any marker (1.4, 16.12) and which can be grasped only by the mind. As J. Fourquet remarks in his Foreword to the *Éléments*, the intra-sentential connections function as the

> articulation of a *lived experience* with the linguistic structure, the structuration of an event for the purpose of linguistic communication.

Connection is the principle of syntax (1.12). It is by means of connections that thought is expressed (1.7); uttering a sentence is to establish connections between lexemes (1.9); and understanding a sentence is to grasp these connections (1.10).

Tesnière strongly insisted that connections are not logical entities. For him, they represent an organic and vital principle of organization (1.8) akin to Humboldt's *innere Sprachform*. The immediate question is: *how to model the connections in a manner compatible with their 'organic' nature*? In our opinion, this is the blind spot of structural syntax, because what is needed is a true modeling and not a mere symbolic notation.

Though Tesnière goes beyond a simple symbolic notation he proposes a *graphical representation* of connections which is still short of a true modeling. He says:

> For greater clarity, we shall represent graphically the connections between words by lines we call *connecting lines*. (1.13).

This move is important. The connections are 'incorporeal' (disembodied) relations which 'externalize' some internal processes of a psychological 'black box'. If we assume they are not logical entities, then we have to represent them differently. But how? With his graphical analogy, Tesnière introduced a basic representation which is neutral relative to the internal structure of the 'black box'. From Tesnière onward there can be as many structural syntaxes as there are paradigms for the 'black box'.

We have already seen that for phonetics there are essentially two major paradigms for the 'black box': the computational one (automata theory) and the catastrophist one (critical phenomena). It is the same thing here: there exist two different types of structural syntaxes. They constitute two divergent developments of the Tesnierian source, each of them moving up in its own way from the graphical analogy of connections to the 'innere Sprachform'. They correspond respectively to the grammatical and actantial dimensions of syntax, whose complementarity was emphasized before.

3.2 *The Stemmas*

Structural connections are directed and hierarchized. They establish dependency relations between the governing and the governed terms (Chap. 2). Whence the notion of *node* (indicating the relation of government between a head and its dependents), and above all, the revolutionary notion of *stemma* (Chap. 3) which appears later in generative-transformational grammars as constituent-structure tree.

This central notion of stemma carries with it a deep ambivalence. On the one hand, Tesnière treats it as a simple tool:

> The stemma shows clearly the hierarchy of connections, revealing schematically the different nodes which unite them into a bundle, and thus allows to concretely visualize the structure of the sentence. (3.9)

But on the other hand, he makes it the basis of the organicity as well as the principle of production of speech:

> The stemma represents the *speech activity* which, under the name of *parole*, is usually opposed to the *result* of this activity as it appears under the tangible and immutable form attributed to a given community and for which is reserved the name of *langue*. This opposition had been clearly understood by Wilhelm von Humboldt who had the inspired intuition for the basic difference between what he refered to by two deeply significant Greek words, *ergon* (langue) and *energeia* (parole). (3.11)

The real issue is to bring together the stemma as 'a visual representation of an abstract notion which is none other than the structural schema of the sentence' (3.10) and the stemma as expression of the linguistic *energeia*. Either we take the stemmas as part of *competence* (*langue, ergon*) and interpret them as syntactic trees generated by rewriting rules, or we take them as part of *performance* (*parole, energeia*) and interpret them as resulting from dynamical processes internal to the 'black box'.

Another consequence of the graphical analogy of connections is to trivialize the problem of the *dimensionality* of the stemmas. The structural order of connections is 'a priori' multidimensional (4.2). Now, though he touches upon this question of multidimensionality, Tesnière confuses the stemma as an objective determination of energeia with the stemma as a tool for representation. He regards the stemma as multidimensional but, to the extent that it is a tree 'constructed as a diagram on a plane surface' (4.5), it is 'necessarily' *two*-dimensional, with respect to the 'graphic possibilities to which it is related' (4.6). Actually, stemmas are colocalization schemas in multidimensional spaces, and there is hardly any evidence that they must be 2-dimensional.

3.3 *The principles of the* Éléments

Starting from the primitive notion of structural order, the *Éléments* proceeds to analyze three main aspects of syntax, viz. connection, junction, and translation. Junction and translation are concerned with the structuring of simple sentences: junction (e.g., connectives) adds two nodes of the same type, while translation allows to change systematically the category of a term (e.g., the participial forms convert verbs into adjectives). As grammatical transformations, translations reflect a hierarchical order among the categories.

Let us see briefly what are Tesnière principles for simple sentences.

1. According to him:

> structural syntax depends entirely on the relations *between the structural order and the linear order.* (6.1)

The multi-dimensional structural order and the linear order are in conflict (deep actantiality vs. grammaticalization) and the resolution of this conflict is, on the one hand, 'the *sine qua non* condition of *parole*' (7.3), and on the other hand, it is one of the main reasons for the typological diversity of languages:

> To *speak* a language is to know what are the structural connections that have to be given up in order to transform the structural order into linear order,

while

> to *understand* a language is to know what are the structural connections not expressed by the sequences, that have to be recaptured in order to transform the linear order into structural order. (7.5)

The direction of the linear order provides a simple *structural* principle for the typology of languages (Chap. 12–14).

2. Viewing the 'inner linguistic form' as the structural schema of sentences enables us to differentiate:

(i) syntax from morphology, which is ultimately concerned with the external phonetic form of sentences (Chap. 15);

(ii) the 'expresser' (the structurally organized signifier) from the 'expressed' (the signified);

(iii) the notion of *meaning* (the relation of the expresser to the expressed), which privileges morphology over syntax, from the notion of *marker* (the relation of the expressed to the expresser), which privileges syntax over morphology.

Syntax is ideal because connections are without markers (Chap. 16).

3. Structural connections define *functions,* that is to say *roles* assigned to words 'as devices for expressing thought' (19.4). Structural syntax is thus also a functional syntax (19.9). For Tesnière there isn't a clear demarcation between the grammatical relations and the semantic roles selected by the actantial relations.

4. Syntax is not only autonomous from morphology, it is also autonomous from semantics. For Tesnière, there is an opposition between syntax, which pertains to the *form* of content, and semantics, which pertains to the *substance* of content (Chap. 20). Structural syntax deals with a 'subjective and unconscious' activity, a 'deep, elementary and necessary' phenomenon (20.13), and not with an 'objective and conscious' activity, a 'surface and purely contingent' phenomenon (20.15). But, though autonomous, syntax and semantics

are evidently strongly correlated since the structural level implies the semantic one and 'there is never a structural connection without a semantic connection' (21.14). But the semantic dependency, that is to say the hierarchical relation between determinant and determinate terms 'works in the reverse direction of the structural connection' (21.7).

5. Functions must not be confused with grammatical categories. Functional syntax is a 'dynamic' one while categorial syntax is a 'static' and taxonomic one (Chap. 25).

3.4 *Verbal node and Valence*

As regards the structure of a simple sentence, Tesnière opts for the *scenic* (actantial) conception of syntax.

> The verbal node [...] expresses *a little drama*. As in a drama, it obligatorily contains a *process*, and very often *actors* and *circumstants*. (48.1)

In other words, an elementary sentence is structurally organized by a verb assigning roles to actants (substantives) and modalized by circumstants (adverbs). We will return (Sec. 4.5) to the division between actants and circumstants. Part of the criticism against Tesnière was due to the fact that it is difficult to understand that, in some cases, spatio-temporal positions function as actants and not as circumstants. Tesnière's fundamental intuition is that the verb is the 'organizing centre' of actantial relations. Structural syntax is a dynamical and *event*-based syntax centered on the *verb*, which is for that very reason opposed to the traditional subject-predicate conception. According to Tesnière, 'we have to see in this conception, a not-yet-discarded *vestige* of the period that extends from Aristotle to Port-Royal, wherein grammar was entirely based on logic' (49.4). If the logical reduction of syntax is being criticized, it is because it masks

> *the interchangeable character of the actants*, which is the basis of the mechanism of active and passive voices (49.14) [...] [and] obscures the theory of *actants* and verbal *valence*. (49.17)

However, while discussing the actants, Tesnière systematically associates the semantic roles with grammatical relations. For him, the first actant is semantically the agent of the action and syntactically the grammatical subject. Similarly, the second actant is semantically the patient and syntactically the direct object. The third actant is semantically the beneficiary or the goal and syntactically the indirect object (Chap. 51). Thus Tesnière postulates a fixed one-to-one correspondance between the three main deep cases (agent, patient, beneficiary) and the three primitive grammatical relations (subject, object, indirect object). In the analysis of the passive diathesis he is therefore led to qualify the agent as the second passive actant (51.27) instead of maintaining it as the first actant (in the semantic sense) and changing the rules of grammatical subjectivization.

We will take up these points further in relation to case grammars and relational grammars. But first, we will return to our critical evaluation of the generativist evidence.

4 The grammatical reduction of structural syntax

Let us first examine the evidences and principles underlying the grammatical reduction of structural syntax in generative grammars.

4.1 *Criticism of the generativist evidence*

4.1.1 *Competence and performance*

We have seen that, according to standard generativism, the task of linguistic theory is to constitute a model of the faculty of language which accounts for the form of humanly accessible grammars. Grammars are generative devices which assign semantic interpretations to phonetic sequences. Syntax is conceived of as a system of formal constraints conditioning such an assignment. In this perspective, a formal grammar must be able to provide each sentence a structural description à la Tesnière. To build generative mechanisms for structural descriptions is the main purpose of syntax, and it is to be complemented by the two semantic and phonetic interpretative components. Such a theory being both descriptive and formal cannot be a theory of linguistic production. It cannot account for the psychological reality of the generative component it formalizes. It is a model of competence and not of performance.

Our main critique against this conception is that it is wrong to bracket the underlying production mechanisms even for the construction of a purely formal theory of syntax. We must seek models which include aspects of performance, even if only in an implicit way, which are able to move up from formalized descriptions of competence to constraints on generating mechanisms (phenomenological abduction), and which treat competence rules as emerging from the underlying dynamics of performance. We propose therefore the following alternative to the generativist 'evidence'.

(i) At the base of the syntactic theory we introduce the speaker's intention, or Humboldt's energeia, that is to say, performance.

(ii) Since a physicalist reductionist theory of performance is beyond our present day technical abilities, we should introduce it only implicitly.

(iii) More precisely, we will consider that the sentence meaning is a global Gestalt describable by an 'internal state' of a psychological 'black box', that is to say by the topology of a complex attractor of an implicit internal dynamics.[#]

(iv) The structural description of the sentence (its syntagmatic constituent structure tree) shows how this attractor catastrophically bifurcates into sub-attractors.

(v) Just as in thermodynamics phase-transition diagrams are rather independent of the fine physico-chemical structure of matter (see the results of the renormalization group), the morphologies of the catastrophic breaking of attractors are largely independent of their topology.

(vi) It is for this very reason that these morphologies can be conceived of as syntactic structures though they are of *semantic* origin.

This key idea was later realized by connectionnist models.

(vii) In this perspective, competence is relatively autonomous, but because the emerging coarse-grained structures are relatively independent of the fine-grained mechanisms they are implemented in. This resolve the tension between the rival syntactic and semantic conceptions of deep structures.

These issues are not mere epistemological generalities. They do have a methodological and programmatic scope. The competence/performance opposition in linguistics is almost parallel to the opposition in physics between mechanics and thermodynamics. The attempts to overcome the latter opposition has been one of the most significant factors in the progress of physics. The interpretation of the macroscopic thermodynamical processes in terms of statistical mechanics at the microscopic level led to major achievements. Linguistics is in the throes of a similar transformation. Its shift towards a 'physics' of meaning requires to introduce in formal 'pure' linguistics the analysis of stable attractors and of their bifurcations. The interpretation of syntagmatic trees in terms of catastrophic breaking of dynamical 'internal states' will achieve a theoretical leap converting their combinatorial algebraic content into a qualitative dynamical one.

4.1.2 *The inadequacy of rewriting rules*

With the sort of 'evidences' we have reviewed, Chomsky *fixed* the eidetic type of grammars and *pre-determined* their formal nature. For example, when he envisages the three main tasks of a linguistic theory to be:

(i) a discovery procedure to induce an adequate grammar from a given corpus;

(ii) a decision procedure to determine the fit of the grammar to data;

(iii) an evaluation procedure to select in a set of possible grammars the optimal one,

and when he claims that:

(i) the search for a discovery procedure is too ambitious;

(ii) the search for an efficient decision procedure is impossible according to well known limitation theorems;

(iii) therefore the search for an evaluation procedure is the goal of a general linguistic theory,

he presumes that grammars of natural languages are necessarily generative in his sense, and that the evaluation problem (iii) has to be taken up only in the framework of this formal class.

But, as we have seen in Sec. 2.2, this way of formalizing grammar is intrisically limited. Though a restricted, local, and short-range generativity is no longer recursive in the logical sense (Sec. 2.2.1), generative theory continues to identify structural descriptions with phrase-markers generated by rewriting rules patterned on the derivation rules familiar in logic. In fact, Chomsky was well aware of this lacuna. He recognized the insufficiency of rewriting rules for generating the structural descriptions, and introduced a transformational component. But this move is not radical enough.

4.1.3 *Remarks on the transformational theory*

The introduction of transformational rules led to what is referred to as the Standard Theory. It consists of:[1]

(a) a base (pre-transformational) component with:
   (i) a categorial sub-component generating a set of elementary initial phrase-markers transformed into kernel sentences by means of obligatory transformational rules;
   (ii) a lexical sub-component responsible for the phonetic and semantic representations, a part of which has also a syntactic function (sub-categorization and selection restriction rules);

(b) a transformational component composed of transformational rules (passivization, negation, interrogation, embedding, etc.) that convert kernel sentences into derived phrase-markers using a series of elementary rules such as expansion (substitution of a single term by a sequence), deletion, addition (e.g., the addition of 'by' in passivization), or permutation.

A sentence is viewed as a set of initial underlying kernel sentences whose phrase-markers define the categories, the functions and the grammatical relations, to which are applied a set of transfor-

1 See for instance Ruwet, 1967.

mations governed by a transformational marker, and resulting in a final derived phrase-marker. If we also consider the assumption that

> only the syntactic information contained in the underlying phrase-markers would be relevant for semantic interpretation of sentences, and similarly only the syntactic information contained in the derived phrase-markers would be relevant for their phonetic interpretation,[1]

then we can distinguish between deep structure (initial underlying phrase-markers to be semantically interpreted) and surface structure (final derived phrase-marker to be phonetically interpreted). The hypothesis that transformations do not change meaning has been severely contested and hence new components were introduced in the syntactic base.

Our main criticism of this conception is that a generativist interpretation of kernel sentences is unecessary. Firstly, in standard theory, the thrust of grammatical recursivity is shifted from the categorial component to the transformational one. Secondly, as we have seen in Sec. 2.2, for a configurational definition of grammatical functions in terms of dependence, i.e., in terms of positions, we only need the stemmatic structure of the phrase-markers, and we are not committed to iterate rewriting rules. Finally, in order to reconstruct the phrase-markers of deep structures from final phrase-markers with deletions or embeddings, standard theory associates with each main category 'an element which has the sole function of representing this category and which has no supplementary lexical specification',[2] and which can be deleted or substituted transformationally. For describing other embeddings, it introduces dummy elements substitutable by relative pronouns. But for all this we do not need generative rules.

In seems therefore that, though notions like phrase-markers and transformations are crucial, it is practically useless and theoretically incorrect to postulate a generative component for the kernel sentences. It seems much more sensible to depend on language acquisition studies and assume that the base consists of a *closed case-structure* component and an *open lexical* component, whose interaction

1 Ibid.: 320.
2 Ibid.: 262–263.

generates the deep structures on which the transformational rules operate.

4.2 *Case grammars*

4.2.1 *The basis for case theory*

Charles Fillmore reintroduced in the late sixties a case conception of grammar in order to overcome the difficulties raised by the transformational-generative standard theory, where, as we have just noted, deep grammatical functions are defined by dominance relations in the phrase-markers of the kernel sentences. Indeed, such a configurational definition does not capture the deep case relations. It is, for example, incapable of accounting for the invariance of semantic roles in passivization, or for the equivalence of the sentences (1) and (2):

(1) John gave a book to Paul,

(2) Paul received a book from John.

In this example a single deep relational structure (sender, receiver, transferred object) is lexicalized by two different verbs specifying opposed selections of the grammatical subject (the sender for 'giving' and the receiver for 'receiving'). As John Anderson pointed out, this way of introducing 'functional' information is not in tune with the standard theory.[1]

These few examples are sufficient to highlight the inherent limitations of projecting surface grammatical relations onto deep structures. Further reflection prompted Fillmore[2] to abandon the base categorial component of the transformational-generative grammar and to adopt case-based grammars: the functional information is no longer introduced configurationally and the functional categories are identified with the deep cases that select the universal semantic roles.

We will examine later in Sec. 5.2 Fillmore's *scenic* conception of syntax. For the moment, we will focus on the basic principles of case grammars.

---

1 See Anderson, 1975a: 21.

2 See the series of papers Fillmore, 1966, 1968, 1969, 1970, 1971a, b, c, 1972.

The founding hypothesis is that there exists a finite list – a closed class – of case universals (or functional categories) whose *notional* content can be determined. These are, for example:[1]

(i) the Agentive (Nominative): the case of the Agent, typically animate, of the state or process described by the verb;

(ii) the Dative: the case of the entity in direction of which the state or process takes place;

(iii) the Instrumental: the case of the force or the inanimate object which causally participates in the state or process;

(iv) the Locative: the case which identifies the place or the spatial orientation of the state or process;

(v) the Objective (Accusative): semantically the most neutral case, the case of the actant whose role in the state or process is defined by the semantic content of the verb.

This founding hypothesis leads to posit that :

(i) a kernel sentence consists of a proposition and a modality (negation, tense, mode, aspect, etc.);

(ii) the proposition is – in line with Tesnière's intuition – constituted of a *verbal node V* assigning a certain number of semantic roles $C_1, C_2, \ldots, C_n$, depending on its *valence;*

(iii) the case C_i (the actantial places) develop into $K_i + NP_i$ where NP_i is a Noun Phrase and K_i a case-marker (e.g., word order, inflection, preposition, adposition, etc.);

(iv) there exist rules of subjectivization and objectivizátion for selecting the deep functions grammaticalized as grammatical subject and object;

(v) there exist linearization rules for word order, which play the role of case-markers (cf. iii);

(vi) lastly, transformational rules apply on the sentences thus formed.

The case conception is an economical and elegant solution to several problems. It explains easily, for example the equivalence of sentences

1 See Fillmore, 1968 and Anderson, 1975a.

(1) and (2) above. In (1) the role of 'John' is determined by the Agentive, that of 'Paul' by the Dative, and that of 'a book' by the Objective. Generally there exists a hierarchy of cases determining the rules (iv) of subjectivization and objectivization. In English, as in many other languages, it is generally the Agent which is subjectivized (and placed in initial position since the word-order is *S-V-O*). Hence we get (1). The passivization subjectivizes the Objective by demoting the Agent to the position of the indirect complement marked by the preposition 'by'. Thus we get (3):

(3) A book was given to Paul by John.

In English, we can also subjectivize the Dative as in (4a):

(4a) Paul was given a book by John.

In French there is no syntactic transformation that allows for subjectivizing the Dative:

(4b) *Paul a été donné un livre par Jean.

But in (2) we use a *lexical* transformation of 'to give' into 'to receive' for subjectivizing the Dative.

Let us note that in (2) the Agent marker is the preposition 'from' (and not 'by') which is also the marker for the local case 'Source' (Ablative). There is a structural equivalence between (2) and (5):

(5) Paul received a book from America.

But it does not imply that in (5) 'America' is an Agent. Indeed the ungrammaticality of (6):

(6) *America gave a book to Paul,

and sentences like (7):

(7) John sent Paul a book from America,

show that the Agentive and the Source cases must be distinguished. We can therefore conclude that in (1) there exists a *syncretism* between Agent and Source. In order to distinguish between these two cases, we can suppose that the Agent must have the semantic feature 'Animate'. If we now compare (8):

(8) John broke the branch,

with (9):

(9) The wind broke the branch,

we have to assign to 'the wind' in (9) the Instrumental case. But notionally, the Instrumental is subordinated to the Agentive, and this fact is marked morphosyntactically by the use of the preposition 'with' as in (10):

(10) John opened the door with the key.

But except by introducing hyper-agents controlling the 'forces' of nature, we cannot say:

(11) *X broke the branch with the wind.

We notice thus a partial interchangeability between the Agentive and the Instrumental, which, as suggested by several comparative studies, appears to be universal (see 4.2.2.3).

The above remarks underline the difficulties encountered in defining the cases purely notionally, making the hypothesis that they univocally determine roles.

However, Fillmore has often insisted on the centrality of syntax. For him, deep cases are 'hidden' functional categories, primarily of the syntactic type, which are to be discovered, delimited and justified by means of *syntactic* criteria. Consider for example (12) and (13):

(12) John polished this table,

(13) John built this table.

These two sentences have the same phrase-markers and apparently the same case-structure (Verb-Agent-Object). However, to the question 'What John did to this table?', we can answer with (12) and not by (13). This syntactic criterion differentiates the Objective case in (12) from the Factitive case in (13). The Objective/Factitive distinction goes back to that between *affectum* and *effectum* in Latin grammars.

This conception of case categories runs into serious difficulties concerning the conflict between the proliferation of cases as syntactic discriminators and their limitation as universals. Fillmore's 'early' theory doesn't solve the problem.[1] If the cases are to be assigned a distinctive function, then the semantics of all the verbs have to be distributed over them, and they proliferate. If instead, they are to be assigned a notional content broad enough so as to drastically limit their list, they become unmanageably broad.

Nonetheless, let us suppose that a restricted number of case universals is available. We can immediately derive from it, on the one hand a first typology of languages, and on the other hand, a rough classification of verbs.

As for the typology, we can immediately explain the well known opposition between *accusative* and *ergative* languages.[2] If we

1 We identify the 'later' Fillmore with the scenic conception of syntax.

2 Fillmore, 1968.

consider sentences involving the Agentive and the Objective of types *V-A*, *V-(A,O)*, and *V-O* we get the following schema for distinguishing the Nominative and the Accusative in accusative languages:

On passivization this schema becomes :

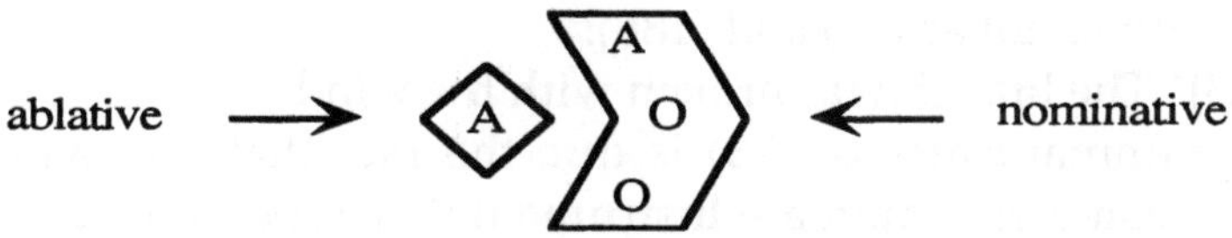

For ergative languages, we have on the contrary the schema:

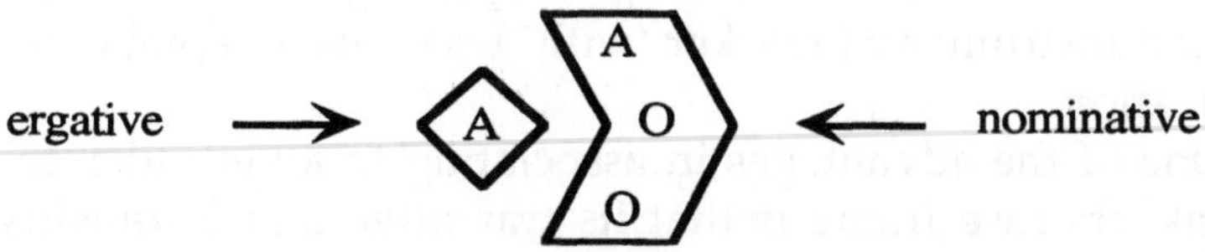

The similarity between the last two schemas explains the widely held belief that ergative languages have the passive voice as their principal diathesis.

As for the lexicon, we assign to each verb a *case schema*. Consider for example the verb 'to cure' in the following six sentences:

(14) Peter was cured;
(15) Peter was cured with this medicine;
(16) This medicine cured Peter;
(17) The doctor cured Peter;
(18a) The doctor cured Peter with this medicine;
(18b) Peter was cured by the doctor with this medecine.

For each of these sentences we can give a corresponding case description:[1] [— O] for (14), [— *O*, *I*] for (15) and (16), [— *O*, *A*] for (17)

---

1 In the following case descriptions the verbal node is omitted and only the case labels are mentioned.

and [— *O*, *A*, *I*] for (18a) and (18b).[1] Assuming the verb 'to cure' has only these six descriptions, we can conclude that it is a trivalent verb requiring the Objective but not necessarily the cases *A* and *I*. Hence the case frame [— *O*, (*A*), (*I*)], where the brackets indicate optionality.

Similarly, consider the verb 'to break':

(19) The branch broke: [— *O*];

(20) = (8) John broke the branch: [— *O*, *A*];

(21) = (9) The wind broke the branch: [— *O*, *I*];

(22) John broke the window with a hammer: [— *O*, *A*, *I*].

It has the same case schema [— *O*, (*A*), (*I*)] as 'to cure'. But one of the differences between the two sets of sentences, in that for (21) there is no such equivalent as (15) and (18b):

(23)* The branch was broken with the wind.

The ungrammaticality of (23) is due the fact that 'the wind' which must be treated as inanimate Instrumental is appearing also as a kind of Agentive. To see that this ambiguity is real, notice the acceptability of:

(24) With such a wind it is not surprising that the branch broke, where the instrumental marker 'with' becomes acceptable to indicate topicalization.

One of the advantages in associating to a verb like 'to cure' or 'to break' its case frame is that its transitive and intransitive forms can be entered in the *same* lexical entry. In this respect, the case grammars are profoundly different from generative grammars. As they employ only configurational definitions of grammatical functions, the latter are faced with a peculiar problem: the subject of the intransitive form has the same selection restriction rules as the object of the transitive form. In general, they solve it by interpreting the transitive form as the causative of the intransitive,

(8) and (20): 'John broke the branch'

being equivalent to

(25) 'John caused the branch to break'.

But it is contrary to linguistic intuition to treat (20) no longer as a kernel sentence, but as a complex one with a nested construction.

Another advantage of the case perspective is that it permits a different kind of semantic grouping of verbs and a new distinction

---

1 *A* = Agentive, *O* = Objective, *I* = Instrumental. "Peter" is treated as Objective. But it could also be treated as the Patient (Dative) or Experiencer.

between synonymy and syntactic distribution. We have seen that 'to give' and 'to receive' are symmetric verbs corresponding to the same prelexical case frame [— *O*, *A*, *D*]; 'to give' requires the subjectivization of the Agentive, and 'to receive' that of the Dative. Similarly, in the pair 'to like' and 'to please', 'to like' subjectivizes the Dative, and 'to please' the Objective. We can also group verbs whose case frames are identical but different relative to the distributional constraints. Such is the case for example with the verbs 'to kill' and 'to murder' (or 'to assassinate'). In 'to kill' the presence of the Dative is obligatory, and so the presence of either the Agentive or the Instrumental (which are mutually exclusive). On the contrary, in 'to murder' the Agentive and the Instrumental are not mutually exclusive, and the presence of the Agentive is obligatory. The case frames are:

'to kill': [— *D*, (*A*)/(*I*)] (where the slash between brackets indicates mutual exclusion),

'to murder': [— *D*, *A*, (*I*)].

Here, we notice that there can be lexical differences of a purely syntactic origin.

The method of case frames also permits for the grouping of verbs that lexicalize different domains of a given semantic field. This is the case, for example, with the verbs of vision (only the obligatory actants are shown):

'to see': [— *O*, *D*],

'to look': [— *O*, *A*],

'to show' (causative of 'to see'): [— *O*, *A*, *D*].

In a famous paper,[1] Fillmore analyzed the verbs of judgment in a similar way. Such analyses are important for they *relativize cases to semantic fields* and thus partially overcome the opposition between cases as semantic universals and cases as syntactic discriminators (see above).

4.2.2 *The difficulties of case grammars*

Like any other theory, case grammars have their own difficulties. Some are local, technical and arise due to the extreme diversity and subtlety of linguistic constructions. Others are more central and highlight some intrinsic limitations of the theory.

---

1 Fillmore, 1971b.

Among the former there is, for example, the problem of equivalence between comitative and coordinate constructions as in (26) and (27):

(26) John came with Paul,

(27) John and Paul came.

There is also the problem of describing the factitive, possessive, and causative constructions as in 'to dream' = 'to have a dream' or 'to suggest' = 'to make a suggestion', etc.

But the examples of the latter kind are evidently far more interesting. We will mention three of them.

a. *Predicative and equative sentences*

Because of their event-based, scenic conception of syntax, wherein verbal valence is a primitive notion, the case grammars cannot account easily for *predicative* sentences like (28):

(28) The sky is blue (or, 'The snow is white' to be more Tarskian),

or *equative* sentences like (29):

(29) The gentleman near the table is the president of our group.

As already mentioned, we notice here a divergence of linguistic conceptions going back to Aristotle and the Stoics. In the verb-based conception, 'to be' in the sense of 'to exist' is a univalent verb, but the copula 'is' in predicative judgments like '*S* is *p*' remains outside the domain of syntax. In order to possess a case description, a sentence like (28) must be transformed into a sentence describing a process like (30):

(30) The sky emits blue.

Conversely, we know that from Aristotle to Port-Royal, the logicist tendency is mainly characterized by the reduction of the verb to a combination of noun and time. By means of the translation Verb→Present Participle, every sentence can be transformed into a predicative one.

In fact, the case and logical dimensions of language are complementary. Even if case grammar is the correct theory of the verb, it should nonetheless be complemented by a correct linguistic theory of predication.

b. The one-to-one selection of roles by cases

One of Fillmore's early hypotheses was that the case functions in a proposition identify univocally the semantic roles. But it cannot be upheld. The semantic roles of the actants in a process are often determined by *several* cases. In (1) 'John' as the Sender actant is both Agentive and Source and 'Paul' as the Receiver actant is both Dative and Target. In (9), 'the wind' is both Agentive and Instrumental. Similarly, if we assume the case frames of 'to see' as [— *O*, *D*] and of 'to look' as [— *O*, *A*], then in (31):

(31) John looked at the car,

'John' will be both Agent and Dative.

To resolve these difficult problems within the framework of a one-to-one selection of roles, we could introduce new cases, for example, a Force case for (9). But this leads to a proliferation of case universals, whose definition depends on the cognitive structure of the world. This is a very delicate issue. Indeed there are very close relations between categories of thought and categories of language. But it is difficult to project an ontology onto the deep structure by way of the notional content of case universals. That is why it seems preferable to conceive of semantic roles as *bundles of case features* (see Sec. 4.3).

c. The ambiguity of case forms

Fillmore's theory interprets deep structures as case relations and ignores the surface morphosyntactic markers of deep cases. Being a semantico-syntactic and not a morphological theory it discards the analysis of case *forms* (for example, prepositions) as pertaining to the peculiarities of a given language. This tendency is justified by the fact that all the theories that sought to assign a notional content to case forms could not cope with the heterogeneity of the notions they encode.

However, as Stanley Starosta pointed out, we should inquire whether this heterogeneity of case content shows structural ambiguities deeper than those resulting from a simple syncretism. A detailed cross-linguistic analysis of languages with very different structures suggests that deep cases do not coalesce (being assigned the same surface case form) at random *and that there does exist some sort of 'universal' case complexes*. Let us mention Starosta's three examples:[1]

(i) the shared distribution of 'with' between Agent and Instrumental is to be found in French, German, Russian, Thai and Tibetan;

(ii) the shared distribution of 'with' among Instrumental, Comitative and 'Complement of manner' (as in 'Peter spoke with passion') is to be found in French, German, Estonian, and Tibetan;

(iii) the shared distribution of 'to' among Dative, Locative and Goal is to be found in French, English, German, Japanese, Korean and Hebrew.

Even if we follow Starosta's 'lexicase' approach which considers case features as lexical, we nevertheless have to pay attention to these regularities because they raise very important theoretical issues. For example, the use of the common marker 'to' for Dative, Locative and Goal shows the ambivalence between an actantial case (Dative) and local cases (Locative and Goal), which is the basis of the localist hypothesis. Similarly, the systematic merging of Agent and Instrumental is due to a drastic limitation of verbal valence and shows that they are two aspects of a deeper case distinguished by the animate/inanimate or intentional/non-intentional oppositions.

4.3 *John Anderson's localism*

After Fillmore, John Anderson tried to tackle the problem of proliferating notional contents of the semantic roles. His position is interesting not only for its theoretical value, but also for its methodology. Anderson applied the structural paradigm to the case category. Just as in phonology or structural semantics units are analyzed in terms of distinctive features, he analyzed the semantic roles into *case features*. The interpretation of deep cases as bundles of features

(i) considerably reduces the number of primitive cases,

(ii) allows to postulate true universals,

1 Starosta, 1975.

(iii) improves upon Fillmore's analysis, and

(iv) explains the case meanings manifested as surface case-markers.

4.3.1 Multicase interpretation of roles

Anderson's point of departure was the observation that semantic roles result from the coalescence of a few primitive cases which can be identified as case-features. By privileging the structure of ergative languages against that of accusative languages, Anderson started off with the Ergative and Absolutive features corresponding respectively to the Agentive and the Objective. Further, on the basis of his 'localist' approach, he chose the Locative feature for cases as varied as the Goal, the Dative , and (of course) the Locative, and the Ablative feature for the Source. These four case features are regarded as primitives, and the rules for the surface subjectivization and objectivization are determined by the distribution of the [Erg] and [Abs] features.

Consider for example the sentences (32) and (33):

(32) = (8) John broke the branch,

(33) John read this book .

In (32) the subject contains only the [Erg] feature, while in (33) it contains both the [Erg] and [Loc] features (since the Dative and Experiencer cases are both interpreted in terms of [Loc]). Similarly, in (34):

(34) = (9) The wind broke the branch,

'the wind' contains the [Erg] and [Instr] features, which explains why passivization marks it by 'by' and not by 'with'.

The multicase interpretation of semantic roles can account for rather subtle differences in interpretation. Consider for example the pair of sentences (35) and (36):

(35) Peter loaded three bags in the car,

(36) Peter loaded the car with three bags.

According to Fillmore, these two sentences will have the same case description. However, we know that they are not synonyms on whereas in (35) the interpretation is 'partitive' (only a part of the car is filled), while it is 'holistic' in (36) (the whole car is filled). For Anderson the partitive/holistic difference can be explained by assigning to 'bags' the Objective case in (35) and the Objective-Instrumental case in (36). Such a description accounts for the differences in passivization as in (37) and (38):

(37) The car was loaded with a bag,

(38) The car was loaded by a bag

For more interesting examples see Anderson (1975a) and (1975b).

4.3.2 *An aspect of the localist hypothesis*

Anderson's theory is a localist one to the extent that it reduces the ambivalence between the local and syntactic uses of cases to a *common principle*. According to it, this evidence long since recognized at the level of case forms, must be transferred to the deep level and become the principle of the deduction of case universals.[1] Anderson retains two semantic axes, namely *locality* for distinguishing [Loc] and [Abl] from [Erg] and [Abs], and *polarity* for distinguishing [Erg] and [Abl] from [Loc] and [Abs].[2]

	Erg	Abl	Loc	Abs
Locality	–	+	+	–
Polarity	+	+	–	–

These substantive universals are deeply rooted in the structure of the phenomenologically experienced natural world.

Such a localist conception leads to reinterpret cases as the Dative. For example consider (39) and (40):

(39) Peter sent a letter from Australia to Canada.

(40) Peter sent a letter to John from Australia.[3]

In (39) 'Peter' carries the [Erg] feature, 'a letter' the [Abs] one, 'Australia' the [Abl] one and 'Canada' the dative [Loc] feature (as Allative, 'to send' being understood as the causative of 'to go'). In (40), on the other hand, 'John' contains not only the [Loc] feature but also the [Erg] one ('to send' being understood as the causative of 'to receive' and no longer of 'to go'). Hence the possibility of Dative transformation by passivization in English, as in (41):

(41) John was sent a letter from Australia by Peter,

1 On this point, see the Introduction of *The Grammar of Case: Towards a Localistic Theory* (Anderson, 1971).

2 Anderson, 1975b: 100.

3 Ibid.: 83 ff.

or by lexical transformation in French, as in (42):

(42) Jean a reçu d'Australie une lettre de Pierre.

This example shows the *differentiation* of *positional proto-actants* into spatio-temporal positions and actants.

Let us compare in the same way the following two sentences (43) and (44):

(43) John bought a book from Peter,

(44) Peter sold a book to John.

At the level of the directional spatio-temporal structure of the exchange, the two sentences have the same structure: 'John' [Loc], 'Peter' [Abl], and 'a book' [Abs]. But, at the actantial level, 'John' contains in addition the feature [Erg] in (43), while it is 'Peter' in (44). The transfer of the [Erg] feature from Locative to Ablative is lexicalized by the verbal duality 'to buy'/'to sell'.

Though simple and limited, these examples show that the deduction of cases must be rooted in archetypal spatio-temporal interactions (locality) interpretable at a second level, either partially or completely, in actantial terms (polarity). In other words, *deep cases must be conceived of as sharing a local base (in the localist sense) constituted of positional configurations with a purely topological and relational content, on which the actantial dimension operates.*

4.3.3 *Localist grammars*

Anderson did not attempt to deepen the status of the localist hypothesis. He took it for granted and was principally interested in deriving from it new grammatical rules.

In his localist grammar, the base has two components. Firstly, a semantic component generating abstract semantic (notional) representations, and secondly a transformational component linking the deep semantic representations to the surface structures (grammaticalization of deep actantiality). The outputs of this base serve as inputs, via the lexicon, to the morpho-phonological component.

According to Anderson, the semantic component is governed by two types of rules:

(i) sub-categorization rules (*SCR*) developing complexes (*C*) of categories and case-features;

(ii) composition rules (*CR*) controlling the composition of the complexes *C*.

In order to account for the transformations (nestings, etc.) Anderson introduced in addition to the initial category *S* (sentence) and the terminal categories Noun (*N*) and Verb (*V*), a pre-terminal category *NP* (noun phrase) and obligatory *CRs* :

CR 1 : *S*→*V*,

CR 2 : *V*→*nom* // *V* (where *nom* represents the neutral case of the Nominative, i.e., the [Abs] feature),

CR 3 : *NP*→*N*.

Some *SCRs* develop the *C* containing *S*, *V*, and *NP* as initial terms. They are of two types:

SCR 1 : χ→±φ (e.g., *V*→± stat(ive) or *V*→± erg(ative)),

SCR 2 : +φ→±ψ (e.g., +erg→± refl(exive), +erg→± caus(ative), {–erg, +caus}→±loc(ative), {–caus, +stat, –refl}→±obl(ique)),

where γ is a category and and φ, ψ case features.

Other *CRs* are of the type:

CR 4 : +φ→φ // *V* (e.g. +stat→cop(ula) // –*V* or +loc→loc // *V*.

(which says that if the feature +φ occurs in a complex then the corresponding category is inserted in a chain of elements immediately dominated by that complex), or of the type:

CR 5: +φ→ψ // γ

(which says that presence of the primitive feature φ in C_0 initializes a *CR* which introduces a secondary feature ψ in the *C* initiated by the category γ and immediately dominated by C_0).

In addition to the rules of the semantic component there are the transformational rules of the syntactic component. Anderson (1971) provides examples of such phrase descriptions based on this blending of case-theoretic and generative conceptions.

4.4 *Relational grammars*

Besides Anderson's localist theory, another very rich approach developed during the seventies was the so-called *Relational Grammars*.

4.4.1 *Basic hypothesis*

The major issue that relational grammars have attempted to tackle is the condition of descent (see, Sec. 2.2.3), that is the characterization of the sub-class of natural grammars within the super-class of formal grammars. For this goal, they identify – as Tesnière did – deep actants with grammatical functions, and postulate that the structure of natural grammars depends upon the three primitive grammatical relations viz., Subject (*S*), Direct Object (*O*) and Indirect Object (*IO*).[1] As David Johnson noted:

> The fundamental tenet of Relational Grammar is that grammatical relations such as 'subject of' and 'direct object of' play a central role in the syntax of natural languages, *i.e.*, they are the proper units for the description of many aspects of clause structure at various derivational levels and figure directly in the statement of numerous grammatical rules and universal principles which govern the structure and organization of the syntax of natural languages. Relational Grammar posits these grammatical relations as primitives in linguistic theory. This contrasts with the position of standard transformational grammar, which views such relations as definable in terms of the constituent structure notions of 'dominance' and 'precedence'.[2]

The base of a Relational Grammar contains a set of primitive grammatical relations $GR = \{S, O, IO\}$, as well as some deep cases. Only the most basic cases (Agent, Objective, Dative) are borne by *GR*s, the others (Instrumental, Locative, Benefactive) being non grammaticalized. The actants assuming these residual cases are called *Oblique Objects* (*OO*). In contrast, the actants supporting the primitive *GR*s are called *terms*, and those supporting the 'nuclear' relations *S* and *DO* nuclear terms.

Kernel sentences are taken as *relational structures* on which are applied transformational rules.[3] Relational structures consist of:

(i) a domain *D* containing a verbal node *V* ('organizing centre' of the relations) and a finite number of *NP*s, $NP_1,...,NP_k$);

---

1 The beginnings of Relational Grammars can be found in Keenan 1972, 1975, 1976, Keenan and Comrie 1972, Comrie 1974, Johnson 1974a, b, c, Postal 1971, 1974a, b. A special volume (1977) of *Syntax and Semantics* series is devoted to them, and a reference work is Perlmutter (ed.) 1983/1984.

2 Johnson, 1977: 153.

3 See, Olmsted Gary, Keenan, 1977.

(ii) a set R_g of *GR*s of type $S(NP_i,V)$, $DO(NP_j,V)$, $IO(NP_k,V)$;

(iii) a set R_s of case relations of the type Inst(NP_i,V), Loc(NP_j,V), etc.

They replace the deep phrase-markers of the standard theory.

Relational grammars are non trivial for, on the basis of cross-linguistic studies, they have been able to demonstrate deep regularities in the *transformations* of relational structures, and have identified particularly interesting *formal universals*. For a brief glimpse of their results we will refer to David Johnson.[1]

4.4.2 *Transformations and Relational hierarchy*

Though non-generative, Relational Grammars are transformational grammars that provide structural descriptions of sentences at several levels of derivation. Transformations can leave the grammatical relations unchanged (as in the case of pronominalization) or, on the contrary (and this is more interesting) transform them. So is the case with:

(i) passivization which promotes the *DO* into *S* ($DO \rightarrow S$) and demotes the initial *S*;

(ii) the dative transformation in English which promotes *IO* into *S* ($IO \rightarrow S$) and demotes the initial *S*;

(iii) Subject-to-Object Raising ($S \rightarrow O$) which transforms into a principal *DO* the *S* of a subordinate clause supporting the *DO* *GR* vis-a-vis the verb of the main clause, as in:

John believes that Bill is ill→John believes Bill to be ill.

An analysis of the transformations show that they consist of elementary operations of promotion, deletion and nesting or embedding. Among the promotion operations are insertions, raisings $S \rightarrow OD$, and promotions $DO \rightarrow S$ and $IO \rightarrow S$ in the passive and dative transformations.

To account for such transformations, we should be able to express what becomes a term NP_1 which is demoted when a promotion assigns its *GR* to another term NP_2. If we couple *GR*s with case-relations, then we can say that NP_1 is demoted of its initial *GR* while maintaining its semantic role. Such is the case with the *S*-Agent in a

1 Johnson, 1977.

passivization where it is remoted of its *GR* of *S* while retaining its Agent role (morphologically marked by the preposition 'by'). Perlmutter and Postal proposed the notion of 'chômage' for the demoted terms and the notation *X* for an 'undetermined' primitive *GR*. In passivization the *DO* is promoted to *S* and, correspondingly, the *NP* supporting the initial *GR S* becomes a 'chômeur' and enters into the relation *X* with the verbal node *V*. Similarly, in the raising (*S*→*DO*) in (iii), 'to be ill' becomes a *DO*-chômeur.

Hence two first principles:[1]

P1: *The Unique Dependency Principle.* When an *NP* assumes a *GR* it can no longer support any other *GR*, and reciprocally a *GR* can be borne only by a single *NP*.

P2: *The Relational Annihilation Principle.* When an *NP* NP_1 assumes a *GR G* borne by another *NP* NP_2 then NP_2 ceases to bear *G*, becomes a *G*-chômeur and acquires the *GR X* with respect to *V*.

Though very simple, this framework can already formulate universal laws *constraining* the possible structure of natural grammars, and this is a step towards the condition of descent. Let us give four such laws:

L1: *The Relational Succession Law.* If an *NP* is promoted by a raising rule, then it inherits the *GR* borne initially by the *NP* it replaces.

L2: *The Agreement Law.* Only terms can trigger verbal agreement.

L3: *The Reflexivization Law.* Only terms can trigger reflexivization.

L4: *The Coreferential Deletion Law.* Only terms can trigger coreferential deletion.

But one of the major achievements of Relational Grammars is to have discovered the *relational hierarchy* of GRs. This is a property of Universal Grammar, in line with Tesnière's deep concept of *translation* (see Sec. 3) as asymmetric (and hence hierarchical) relation between categories.

---

1 See Johnson, 1977: 155.

L5: *Law of relational hierarchy*. *GR*s are hierarchized according to the order

$$RH = S > DO > IO > OO.$$

The validity of *RH* can be tested at different levels. Firstly, as Edward Keenan and Bernard Comrie showed in a key paper, it governs the *accessibility* of relative clauses.[1]

L6: *Accessibility to Relative Clause Formation*. If a language can relativize an *NP* holding the position *P* in the accessibility hierarchy

$AH = S > DO > IO > OO > NP$ of possession > comparison O,

then it can relativize the *NP*s holding a position higher than *P* in *AH*.

Further, it governs the general law of increasing rank discovered by Perlmutter and Postal. Within a relational structure, a constituent *A* is of a higher rank than a constituent *B* if *A* governs *B* unilaterally, or if *A* and *B* reciprocally govern each other and $A > B$ in *RH*.

L7: *The Reranking Law* (Perlmutter and Postal). Every rule that transforms the *GR* of a term increases its rank. In other words, every rule that transforms the *GR* of a term raises it in *RH*, or raises it, by disembedding, into a hierarchically superior sentence-constituent (for example, subject-to-object raising).

A special case of this universal law is Johnson's law of promotion into subject position, which states that the $S \to P$ promotions constitute a *relational chain*.

L8: *The Advancement-To-Subject Chaining Constraint* (*S-promotion Law*). If a language can promote an *NP* of position *B* in *RH* into the subject-position *S*, then it can similarly promote the *NP*s of positions $A > B$ in *RH*.

This law is clear from the comparative table below:

French	DO +, IO –, OO –

1 See, Keenan, Comrie, 1972.

German	DO +, IO –, OO –
Albanese	DO +, IO –, OO –
Japanese	DO +, IO +, OO –
Sanskrit	DO +, IO +, OO –
Malagasy	DO +, IO +, OO +

Similarly, it seems that promotions *S*→*DO* also give rise to a relational chain, and hence are governed by the following law.

L9: *The Advancement-To-DO Chaining Constraint* (*DO-promotion Law*). If a language can promote oblique objects *OO* (Instrumental, Benefactive, Comitative) into the *DO* position, then it can do it for indirect objects.

This happens, for example in English with double accusative constructs. English has e.g., promotions Benefactive→*DO* (John bought the book for Mary→John bought Mary the book), and it also has *IO*→*DO* promotions (John gave the book to Mary→John gave Mary the book).

One can thus propose an exclusion principle between promotions from the same source but with different targets:

P3: *The Target Uniqueness Principle* (*Exclusion Principle*). There cannot exist in a language two promotion rules $X_i \rightarrow Y_i$ and $X_j \rightarrow Y_j$ with the same source ($X_i = X_j$) but with different targets ($Y_i \neq Y_j$).

Together with laws *L8* and *L9*, this principle makes relational grammars predictive (and therefore refutable). It predicts, for example, that if, as in Japanese, a language can subjectivize indirect objects (*IO*→*S*) then it cannot objectivize the oblique objects (*OO*→*DO*). Indeed, as per *L9*, *OO*→*DO* implies *IO*→*DO*. But, *IO*→*S* and *IO*→*DO* are mutually exclusive according to *P3*.

Although this approach to Universal Grammar is both elegant and economical, it raises delicate questions. One example is the *impersonal passive* in languages like Spanish, Latin, German, Danish, Polish, Welsh or Finnish. It has been studied by B. Comrie.[1] In general, passivization consists of a promotion *DO*→*S* coupled with a

1 Comrie, 1977.

demotion (even a deletion) $S \rightarrow X$. But in languages with an impersonal passivization, there exists a *spontaneous* demotion $S \rightarrow X$ which is not induced by any promotion $DO \rightarrow S$. This phenomenon challenges certain principles as *P*2 making demotion an obligatory consequence of a correlative promotion. In Spanish and Polish, the spontaneous demotion is necessarily a deletion and we can assume that it is an impersonal subject which induces the demotion of the initial *S*. But in Welsh this is not the case, and the Reranking law (*L7*) is violated.

Another problem is that of deciding whether the hierarchies *RH* and *AH* are really universal. They do not seem to be. Firstly, several languages treat oblique objects *OO* as direct objects *DO* by nesting. In many East African and South East Asian languages (Vietnamese, Thai, etc.) a sentence like (45):

(45) John kills the chicken with a knife,

will be expressed as (46):

(46) John uses a knife (for) killing the chicken.

Furthermore, certain languages may be lacking some primitive *GR*s and in particular *IO*. So is the case in Kinyarwanda (Bantu) language as attested by Judith Olmsted Gary and Edward Keenan. Such examples lead to a refinement of the Relational Hierarchy law (*L5*):

(i) *RH* may be incomplete,

(ii) the same *GR* (and not just *OO*) may be supported by several *NP*s (double *DO*, double *IO*, etc.),

(iii) the rules may apply only on subcategories of *GR*.

There are several instances of (iii). The *DO*s of stative verbs cannot in general be promoted by passivization:

(47) This work costs some energy,

(48) *Some energy is cost by this work.

In English, the dative transformation $IO \rightarrow DO$ applies only on Datives of transfer verbs (give, say, sell, show, etc.):

(49) John added a book to the pile.

(50) *John added the pile a book.

In Kinyarwanda, the *DO* is characterized as *GR* by standard criteria such as:

(i) Subject-Verb agreement;

(ii) position (*SVO* order) (as for *S*, but not for *OO*);

(iii) non-marking of cases (as for *S*, but not for *OO*);

(iv) pronominalization (as for *S*, but not for *OO*);

(v) reflexivization (not for *OO*);

(vi) passivization (not for *OO*);

(vii) relativization (as for *S*, but not for *OO*).

But the *IO* (Datives of transfer verbs) and the Benefactive present exactly the same characteristics as *OD*, and there are thus sentences with *double DO*. Evidently, the Dative exists as *locative* and it is then marked by a preposition.

The sentence (51):

(51) John sent a letter to Mary,

will be rendered by a sentence with a double *DO*. If it is rendered with the equivalent of the preposition 'to', the meaning changes: 'Mary' will no longer be the Receiver, but her spatio-temporal localization (52):

(52) John sent a letter to Mary's house.[1]

Similarly, a sentence with Benefactive:

(53) John writes for Mary,

will be rendered by a sentence with a double *DO*. If 'Mary' is marked by the preposition 'for' then 'Mary' will no longer be a Beneficiary but a *causative* actant, (53) having the meaning of (54):

(54) John writes because of Mary.[2]

The hypothesis regarding relational structures with double *DO* can be verified with respect to several other properties of Kinyarwanda and thus prompts us, contrary to the Uniqueness Principle (*P*1), to assign a *valence* to *GR*s.

Irrespective of the difficulties they involve, relational grammars provide an excellent framework for a Universal Grammar. The *RH* specifies 'quasi'-universal laws governing the rules of promotion and raising. But it seems also to govern the rules of reflexivization and coreference deletion.[3] If reflexivization is possible between *A* > *B* (and

1 This is a fine illustration of the localist hypothesis.

2 See, Olmsted Gary, Keenan, 1977: 107.

3 Coreference deletion refers to the deletion in an embedded clause of a pronoun that anaphorically replaces an actant of the main clause. Examples: 'The hen is good for eating (it)', 'The sea-lions are funny to watch (them)', 'He

then A is necessarily the NP and B the reflexive pronoun), it is possible for the whole segment SB in AH. Similarly, if coreference deletion is possible for A/B, then it is possible for $S \geq P \geq A/S \geq Q \geq B$. These two rules form Ross' general principle:[1]

P4: *The Primacy of Subject Principle* (*the Primary Constraint*). Subjects have precedence over objects and no rule can be applied more freely on objects than on subjects.

4.4.3 *Grammatical relations and actantial schemas*

As we have just seen, the relational grammars formulate principles and laws which considerably *constrain the arbitrariness* of linguistic descriptions. In this sense, they gain an explanatory power. Starting with primitive *GRs*, they focus on the relational hierarchy *RH* and develop the idea that grammatical rules operate on *continuous segments* of it. As pointed out by Johnson,[2] their results can be abstractly presented as follows.

The transformational rules are in general of the type: $G_i(NP,V) \rightarrow G_f(NP,V)$ where G_i is an initial *GR* and G_f the final *GR*. Such a rule will have a lower limit *Inf* and an upper limit *Sup* for any G_i. The two principles are:

P5: If $G_f = S$ or DO then *Sup* is a universal (i.e., it is not language-dependent).

P6: $G_i \rightarrow G_f$ is valid for every $Inf \leq G_i \leq Sup$.

But it is indeed surprising that, after having recognized the importance of *GRs* as primitives, Relational grammars did not attempt to understand their *origin*, and to *deduce* them from general principles. They treated *GRs* as *abstract* relations $G(NP,V)$ without noticing that such a treatment is fundamentally insufficient.

---

is driving the car for checking (it)', etc. But, *'John wished that Mary kiss (him)'.

1 Ross, 1974.

2 Johnson, 1977: 173.

4.5 *The symbolic misunderstanding of valence*

Our investigations reveal a sort of blind spot in structural syntax, concerning the true nature of the theoretical categories of connection and valence, which are substituted either by a graphic drawing or by a logical symbolization. Structural syntax should make a qualitative leap and frontally tackle the problem of a *deduction* of cases. But this imperative is generally misunderstood. If we look, for example, at the Proceedings of the Conference on *Valence, Semantics case and Grammatical Relations*,[1] it is very striking to observe the systematic logicist underestimation of the valence problem. Syntactic relations are treated as formal relations $R(x_1, \ldots, x_n)$ in the framework of categorial grammars (in the Polish sense later improved by Montague). This is for instance the case for the valence theory proposed by Hartmut Günther.[2]

As is well known, categorial grammars are based on the guiding principle that in natural languages there are many more categories than the usual ones (noun, verb, adjective), and that it is possible to develop a formal calculus of categories as in logical type theories. In Günther's simplified version of Montague grammar, one starts with the basic categories V (sentence radical, verbal node), N (nominal syntagm) and C (common noun). Complex categories are then derived recursively.

Principle: If A and B are categories, then A/B is a category, and if $b \in B$ and $\alpha \in A/B$ then $\alpha(b) \in A$ (where $a \in A$ means that the element a is an expression of type A).[3]

From this principle it is easy to deduce a recursive definition of the actants of an expression. If $\alpha \in A/B$ where $A \neq B$, we say that $b \in B$ is a *complement* of α in the expression $\alpha(b) \in A$. (If not, we say that α is an *attribute* of b. If $A = C$, an adjective $\alpha \in C/C$ is a typical case of an attribute.) We now say that b is an *actant* of an expression α if b is a complement of α or of $\gamma = \alpha$ + complement of α, etc., recursively.

1 VSG, 1978.
2 Günther, 1978.
3 A/B is like the type of maps from B to A.

The sentence, 'Peter gave the book to John' will for example have the structural description:

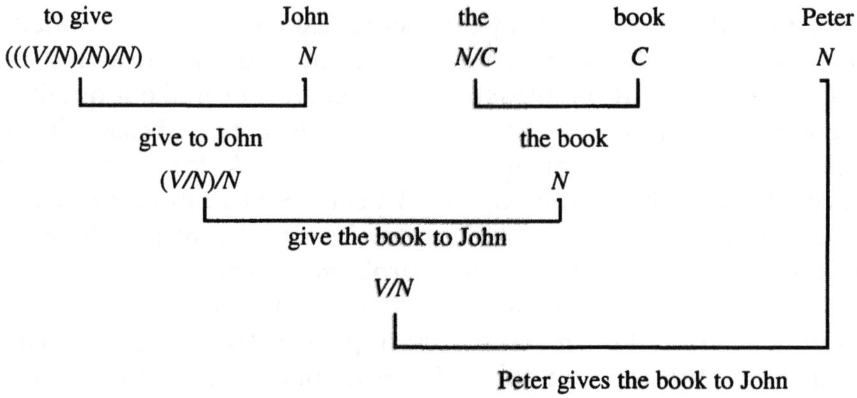

Whatever was the intention of its author, this categorial definition of valence remains insufficient. The valence of a verb is not reducible to the number of its arguments. What requires explanation, is the radical limit of valence in natural languages. It has nothing to do with a purely psychological limitation of formal complexity, and is impossible to understand in a formal framework wich does not satisfy the condition of descent. When Günther claims:

> The present definition of actants and valence is made in purely syntactic terms. The syntax used is, however, a categorial syntax, i.e., a semantically based syntax. This cuts us off from any dispute as to whether a certain property or certain expression is syntactic or semantic. We have a formalism, and if a linguistic element shows a certain property when described within this formalism, it has syntactic valence.[1]

he is theoretically too optimistic. The internal structure of valence raises the problem of the *form* of relational morphologies from where we must deduce a *configurational* definition of semantic roles.

The question of the schematization of deep actantiality is that of a geometric figure-like representation of syntactic relations, which can neither be a graphic drawing nor a symbolic notation but rather an objective determination of the connections. Since Tesnière, it has been adequately posed only by Charles Fillmore:

---

1 Günther, 1978: 151.

> The main problem is how one can indicate the case role of noun-phrases and embedded clauses in the sentences of which they are constituents, and what consequences the choice of notation has for the operation of the grammar.[1]

Fillmore marked the case roles by labelled nodes dominating the *NPs*, even though cases are not categories. But it was a temporary solution. He was at the same time recommending a stemmatic notation.

> I have in mind a kind of dependency notation which makes use of kernel trees or 'stemmas' each containing one root-node, one or more labeled branches, and a variable or index symbol at the leaf end of each branch. The node is a complex symbol containing semantic, phonological and rule-features information, as well as the case valence. The branches are labeled with case labels and are ordered from left to right according to the case hierarchy. The variables at the leaf end of the branches represent the entities which bear case relations to the predicator represented at the node. Any sentence has at base a collection of stemmas of this type, plus information about identities involving the variables; either there can be co-reference among the variables, or some of the variables can be identified with some of the stemmas. That much identifies the semantic interpretation of the sentence.[2]

According to this view, a sentence would be constituted basically of a collection of stemmas and of a building plan governing their groupings and embeddings (inter-stemmatic identification of variables, identification of a stemma with a variable of another stemma in the case of embedding, etc.). The stemmas organized in this way would then serve as inputs to the transformational cycle.

5 The scenic conception of case roles and the morphodynamical schematism

5.1 *The distribution of actantial semantism*

As we have seen in Sec. 4.2, one of the central difficulties of case grammars is to balance the delimitation of cases imposed by their

1 Fillmore, 1971b: 53.
2 Ibid.: 55.

universality and their proliferation imposed by their syntactic function and their dependence on the semiotic-cognitive organization of the natural world.

As pointed out by Stefan Fink, case grammars deal with 'semantic structures that act as filters in the transformation of thoughts into language'[1] and should thus account for the fact that

> the organization of human knowledge about the world at large is based on certain fundamental relationships, such as causation, time, space, etc.[2]

But at the same time, the stemmas which serve as inputs to the transformational cycle are not semantico-cognitive representations. They are purely syntactic relational structures. If we wish to unify these two contrary requirements in a theory where cases are conceived of as substantial universals, defined once for all and globally by a notional content, then as emphasized by Dominique Willems, the case roles become too abstract. They 'end up losing any semantic value'[3] and, at the deep level, the theory slips back into the vicious circle that it denounced in the traditional theories which attempted to define surface cases in purely notional terms.

In Sec. 4.3 we sketched the principles of Anderson's solution, which:

(i) treats case contents as complex meanings analyzable into case features;

(ii) identifies, on the basis of the localist hypothesis, case features with a restricted number of universals;

(iii) assumes that verbs select case features;

(iv) elaborates a 'generative' grammar for these selections.

The idea of a 'generative' case system is very attractive. But it remains nevertheless insufficient to the extent that it is not able to understand the origin and to warrant the deduction of case features. The locality and polarity primitives remain unquestioned. Their spatial and dynamical nature is certainly posited but only in terms of a *semiotics* of space, and not of a *geometry*, while the problem is precisely that of

1 Fink, 1978: 180.
2 Ibid.
3 Willems, 1978: 247.

reducing the case semantism – especially its localist dimension – to a geometrical configurational determination.

We will return in Sec. 6 to this internal contradiction of the *semiotization* of the localist hypothesis. In this section, we will present two ideas permitting a reduction of the case roles to pure positional values defined by a schematic figuration of connections. The first idea is to move from autonomous and independent cases defined by notional labels to those defined by the internal relational structure of the *verbal nodes*. The second idea is to take into account the fact that *several* different semantic fields can be the underlying substratum of a *single* schema of actantial interaction.

Consider for example the verbs 'to give' and 'to go'. Their common schema is a schema of *transfer* between positional proto-actants (places), namely a transfer of an abstract place T from a source place S to a goal place G. But this schema operates on two different substrata, its purely localist structure interfering in two different ways with the dimensions of intentionality, and agenthood. In 'to go' the source S and the goal G are pure spatio-temporal localizations and T is an Agent. On the contrary, in 'to give' the source S is a Sender, the goal G an animate (anthropomorphic) goal-actant (Addressee), and T an Object. We see that *if the principle of identity of the actants is reduced to their localization* 'to give' and 'to go' become isomorphic schemas: *actantial schemas are equivalence classes of relational structures which become isomorphic when their actants are reduced to pure places.* We will call 'proto-actants' such reduced actants.

This elementary example shows why it is necessary to bring under a common positional base the 'local' and the 'grammatical' instances of deep cases. 'To give' and 'to go' have in common the transfer schema which is a *syntactic* schema of a purely *topological* nature. The only thing that distinguishes them is the manner in which the positional proto-actants are specialized into actants (A) or locations (L). If we use a rough graphic representation $S \xrightarrow{T} G$, we could give the following structural descriptions for 'to give' and 'to go':

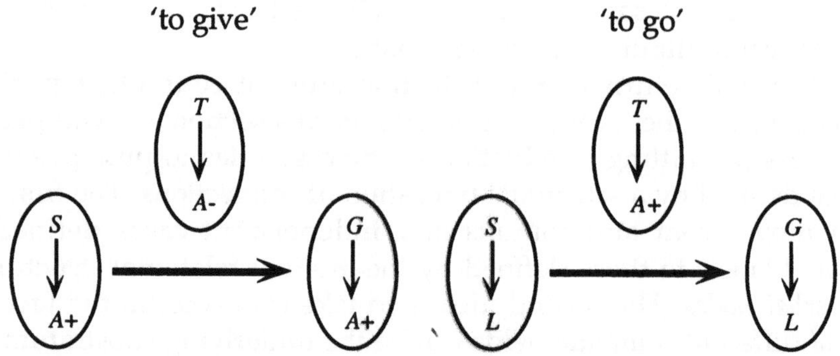

where the arrows inside the blobs indicate the relation of actantial specialization: $A+$ the animate actants (actors), $A-$ the inanimate ones (objects), and L the locations. In these descriptions, the cases are defined *configurationally*:

(i) as a function of their position in the schema;

(ii) as a function of the specialization $X \rightarrow (A,L)$ of place X into an actant or a location;

(iii) as a function of the context:

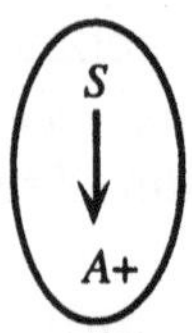

is the Sender as sub-categorization of the Agent,

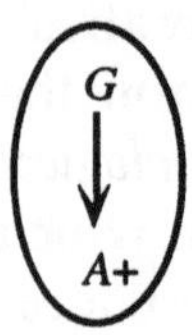

is the Addressee as sub-categorization of the Dative,

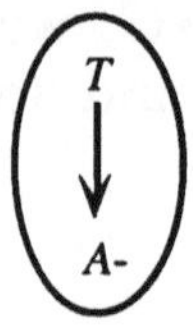

is the Object in the context 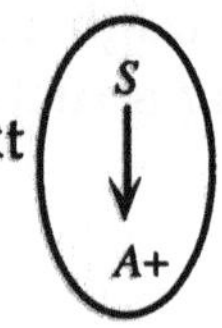 and 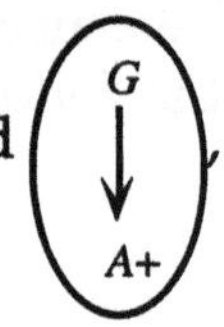,

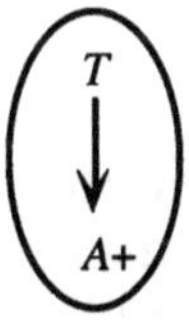

is the Agent in the context 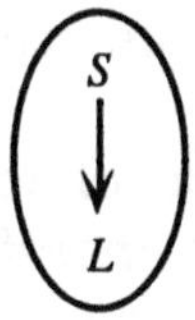

and 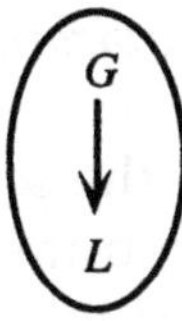

, etc.

Though too rudimentary, this example shows that if one wants to define a fixed and global notional content of cases (Agent, Dative, etc.) then one is obliged to confuse positional values as different morphologically as $S \to A$ and $T \to A$.

We may also observe that in this description, the interpretation of 'to give' as a sort of factitive of 'to go' involves a transformation of specializations of the proto-actants: $A+ \to L$ for S, $A- \to A+$ for T, $A+ \to L$ for G. And finally, we notice that other verbs of transfer are distinguished from 'to give' and 'to go' by other specializations or by the absence of certain specializations:

'to send' 'to receive'

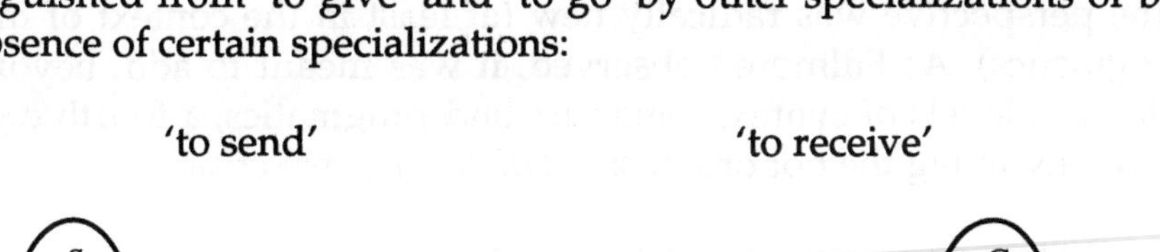

We can then separate in the verbal semantism what belongs to the schemas and what belongs to the semantic fields on which they operate. As suggested by Timothy Potts:

> Wherever there are isomorphisms between sets of interrelationships we should, I suggest, recognize only one set of semantic roles. The difference between the roles in the two situations will be accounted for by the different semantic fields in which they lie, the semantic field working upon the role to modify it.[1]

1 Potts, 1978: 454. See also, Fink, 1978.

5.2 *The relativization of case roles to scenes*

The idea that the schemas of actantial interaction operate on different semantic substrata, and therefore that the notional content of case roles is *contextual,* has been deeply developed by Fillmore, particularly in his key paper 'The Case for case reopened.'[1] Fillmore begin from the assumption that there are *verb classes* (such as verbs of judgement, of movement, etc.) associated with different semantic fields which he refers to as *scenes*. Each scene is lexico-syntactically organized by a limited number of specific constructions which select the respective cases 'having specific syntactic and lexical features.'[2] Hence the celebrated slogan:

> Meanings are relativized to scenes.[3]

The perspective was radically new (at least in the context of modern linguistics). As Fillmore [4] observed, it was meant to add, beyond the classical levels of syntax, semantics and pragmatics, a fourth *cognitive* level involving the notions of *orientation and perspective*:

> My current position is that it is the orientational or perspectival structuring of a message which provides the subject matter for the theory of cases, and that the case notion figures very differently in grammatical description from what I originally had in mind.[5]

In order to develop his scenic conception, Fillmore took recourse to the notion of *case-frame* mediating between the description of situations and the underlying syntactic representations. The case-frame assigns semantico-syntactic roles to the actants of the process expressed by the sentence and this assignment constraints the choice of a perspective that selects one of the actants as subject according to a *case hierarchy*. Fillmore maintains a *conceptual* (i.e., cognitive, prelinguistic) definition of cases, introducing the key idea of a *'pictorial'*

1 Fillmore, 1977.
2 Willems, 1978: 247.
3 Fillmore, 1977: 59.
4 Ibid.: 60.
5 Ibid.: 61.

similarity between the syntactic structure of a sentence and the scene (the states of affairs) it describes:

> Such descriptions [are] in some sense intuitively relatable to the way people think about the experiences and event that they [are] able to express in the sentences of their language.[1]

Nevertheless, case universals are not however ontological categories. They are forms of relations. But these forms reflect the forms of events in the phenomenological world. They are abstract, but conditioned by the same spatio-temporal a priori as natural phenomena.

In 'The Case for case reopened', Fillmore explains how he developed the scenic conception in response to the criticisms of his 'The Case for case'. He addresses a number of issues in order to justify his emphasis on schematization. He begins with Katz' and Chomsky's criticism that case grammars are mere notational variants of the Standard Theory. He points out that, even if paradigms can be translated one into another, the value of a paradigm is measured by the relevance of the questions it raises when applied. Now, contrary to the generativist paradigm,

> a strong assumption about the deep structure of cases forces the analyst to ask certain questions about the number and variety of the semantic functions of the parts of sentences.[2]

Fillmore goes on to state that the concepts associated with cases are very difficult to define and that their definitions differ according to the linguists (see Sec. 4.1). That is the main difficulty:

> The next truly worrisome criticism of case theory is the observation that nobody working within the various versions of grammars with 'cases' has come up with a principled way of defining cases, or principled procedures for determining how many cases there are, or for determining when you are faced with two cases that happen to have something in common as opposed to one case that has two variants.[3]

1 Ibid.: 62. See, sec. 1.2.4 for the correlation between logico-syntactic structures and states of affairs.

2 Ibid.: 67–68.

3 Ibid.: 70.

The word 'principled' is of course crucial. It invokes what we have been referring to as the problem of case-deduction.

Relativization of cases to scenes involves defining semantic roles on the basis of prototypical situations. Consider for example, the prototypical situation of commercial exchange. It consists of typical actants like 'seller', 'buyer', 'money' and the 'object bought', and it refers to a schema of *double transfer*. Fillmore's guiding ideas are the following:

(i) Every sentence describing this prototypical process brings it into perspective in a particular manner:

> A prototypical commercial event involves all these things, but any single clause that we construct in talking about such an event requires us to choose one particular perspective on the event.[1]

The choice of a perspective is manifested lexico-syntactically in the choice of the verb and in the selection of grammatical relations. If the perspective focuses on the seller, then one will choose the verb 'to sell'; if it focuses on the buyer, then one will choose 'to buy'; if it focuses on the seller-buyer-money relation, then one will choose 'to pay'; if it is focuses on the money-object relation, then one will choose 'to cost'; etc. In other words, for Fillmore, every linguistic expression will involve certain types of topicalization and focalization.

(ii) Every verb linked to the situation of commercial exchange activates its scene *globally*. This basic presupposition transforms the nature of the theory.

> The new question for the theory of cases is this: What do we need to know about the various participant roles in a situation in order to know which of these roles or which combinations of them can be put into perspective, and then, for those which have been put into perspective, which is to become the subject and which is to become the direct object?[2]

These two very natural ideas are deep and programmatic. They are deep because they redefine the role of semantics in structural

1 Ibid.: 72.
2 Ibid.: 73.

syntax. 'The study of semantics is the study of the cognitive scenes that are created or activated by utterances'.[1] We select and understand linguistic expressions by 'exciting' in our minds scenes, memorized prototypical situations, in relation to which the expression has a nominative, descriptive or classificatory function. In other words, while perspectivizing, an expression invokes the global background on which it is profiled. As Fillmore notes: 'It is as if descriptions of the meaning of elements must identify simultaneously 'figure' and 'ground''.[2] Scenes are Gestalten, and semantics is as a high level abstract perception.

These two ideas are also programmatic in that they redefine what can be the base of a grammar.

(i) A scene Σ consists of:

(a) a semantic isotopy I (e.g., 'commercial' in the commercial exchange scene);

(b) a *global schema* G of interaction between positional proto-actants P_i, defined in an underlying abstract space Λ;

(c) specializations of proto-actants P_i into actants (actors, objects, forces, etc.) or locations.

(ii) Σ defines the case-roles of the process at the same time *configurationally* (because of (b)), *semantically* (because of (a)), and *notionally* (because of (c)).

(iii) Generally Σ, when linguistically expressed, will be embedded in space-time $\mathbb{R}^4$ via a map $j : \Lambda \rightarrow \mathbb{R}^4$. Through j, the actants specialized into locations become spatio-temporal actants and actants specialized into actors or objects become localized. Localization is linguistically manifested through circumstants.

(iv) There exists a limited number of archetypical universal localist schemas $\Gamma_1, \ldots, \Gamma_k$ determining the case universals.

(v) To perspectivize Σ, an expression has to *cover* the global schema G, either partially or globally, by archetypal local schemas.

---

1 Ibid.: 73.

2 Ibid.: 74.

(vi) In general, there will be several ways of *gluing* archetypes Γ_i to cover G. Gluing operators act as *anaphors*.

(vii) The choice of an archetype Γ_i, that is, of a 'morphism' $h_i : \Gamma_i \rightarrow G$ between the archetypal (universal) local schema Γ_i and the global (particular) schema G is manifested, via the semantic isotopy I (i–a)), by the choice of a verb V_i (to sell, buy, pay, cost, etc.). Through its 'commercial' feature V_i 'excites' globally Σ. But through its valence, it acts as a verb of type Γ_i.

(viii) What Fillmore calls the *saliency hierarchy* determines the minimal part of G that *must* be covered in order to express Σ correctly.

(ix) A *case hierarchy* determines the way in which the actants of the morphisms $h_i : \Gamma_i \rightarrow G$ are taken over by grammatical relations.

(x) The part of G that is not covered by archetypes can be expressed by other sentences (with anaphorization, see (vi)) or by adverbs, subordinate clauses, etc.

(xi) Once grammaticalized, the kernel sentences associated with the morphisms h_i serve as inputs to various transformational cycles.

5.3 *Towards the localist hypothesis*

The topological and relational interpretation of Fillmore's scenic conception shows that it is mistaken to attribute autonomous notional contents to cases. If one merges the cases relative to the scenes into 'archi-cases' such as Agent, Objective, Beneficiary, etc., then one slips back into the vicious circle opposing their limitation as universals to their proliferation as clause discriminators (see Sec. 4.2). Cases are *types of relations* semantically relativized to scenes and if one can speak of them as universals, *it is not at the level of the notional archi-cases, but only at the level of positional proto-cases.*

The problem is however more complex. We can in fact apply to the local archetypes Γ_i the specialization of the positional proto-actants into actors, objects, locations, etc. We get then what may be called *archetypal microscenes* endowed with a *canonical semantics*. If we consider, for example, the transfer scheme

$$S \xrightarrow{T} G$$

of Sec. 5.1, the specialization of S and G into actors (Agent-Source and Receiver) and T into an object is *canonically* lexicalized by the verb 'to give' (or by its synonyms), and the specialization of S and G into locations, and T into an actor (Agent) is canonically lexicalized by the verb 'to go' (or by its synonyms). This *primitiveness* of 'to give' is lexically manifested in the fact that in a scene like the commercial exchange, one can transfer the semantic isotopy I onto one of the actants (using for instance the term 'money' = object + trade) and replace 'to pay' by 'to give', etc. The primitiveness of verbs associated with the canonical semantics of archetypal micro-scenes is essential to syntactic theory.[#] It leads to the critical problem of a deduction of the universals Γ_i.

Assuming this basic problem resolved, one could develop the scenic conception according to the 11 points presented in Sec. 5.2, reformulate relational grammars as well as transformational grammars, and elaborate a theory of universal grammar.

But for this, it is necessary to undertake a qualitative leap, and it is here Catastrophe Theory (CT) comes into the picture.

5.4 *Structural Syntax and Catastrophes*

CT proposed the first morphodynamical schematization of actantial relations. We will sketch it, but only very briefly because our presentation will add nothing to what René Thom himself developed in *Stabilité Structurelle et Morphogenèse* and *Modèles Mathématiques de la Morphogenèse*, and to what has been further elaborated by Wolfgang Wildgen in *Catastrophic Theoretic Semantics*.[1] Moreover, we will be extensively dealing with the narrative actantial interpretation of the elementary catastrophes in another book.[#]

Given a spatio-temporal process involving actants, Thom associates to it an abstract structure that is *both objective and syntactic.*

\# We think that the best framework for developing this idea is the concept of *blending* introduced in the mid nineties by Gilles Fauconnier and worked out by Rick Grush and Nili Mandelblit.

1 Wildgen, 1982.

\# See *Physique du Sens*, Petitot, 1992.

This 'common root' between objectivity and syntax is called an *actantial graph* and is derived from the reduction of actants to their locations.

To take a standard example, let us consider a process of 'capture' of an actant S_2 by an actant S_1. The corresponding graph is presented in fig. 13.

Let us make some observations on this graph which might appear trivial.

(i) The space-time involved in fig. 13 is not the physical, global space-time endowed with its invariance group, but only a local chart Λ endowed with a 'poor' (differentiable) structure. Λ is embedded in $\mathbb{R}^4$ by the embedding $j : \Lambda \rightarrow \mathbb{R}^4$ mentioned earlier. Spatio-temporal localization corresponds to j. Localization in Λ is of a completely different kind; it concerns the *immanent* spatio-temporality of the process and not its positioning relative to a coordinate frame in external (transcendent) space-time $\mathbb{R}^4$.

(ii) The 'capture' graph is objective, but in a very qualitative, coarse-grained, sense. All the physical specificities of the process are bracketed and the actantial interaction is reduced to its pure catastrophic infrastructure. It is this reduction that permits to shift from the objective to the syntactic level.

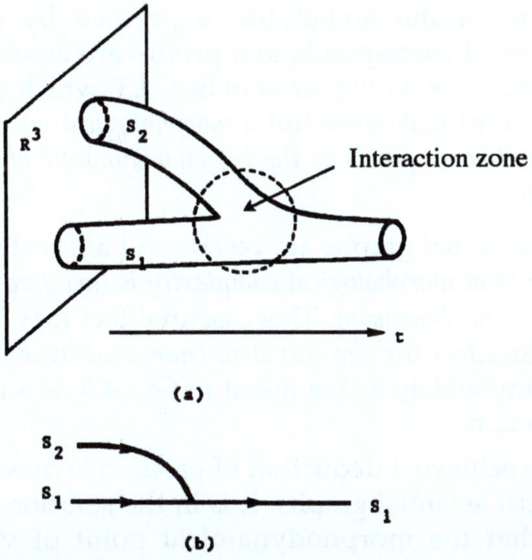

Fig. 13: The actantial graph of capture. (a) The temporal evolution of the locations S_1 and S_2 and the zone of interaction. (b) Reduction of the capture process to its actantial graph: the arrows symbolize the permanence of the actants and the vortex the interaction.

(iii) The kernel sentence describing the capture process is syntactically isomorphic with its actantial graph. Its deep case structure includes a verbal node, an Agent, and an Object. According to the subjectivization and objectivization rules, it will be grammaticalized by a transitive sentence *SVO*.

(iv) The cases of the actants are defined here purely *configurationally*. As Thom explains: 'the subject [the Agent] is the actant that survives the first catastrophe of the process, symbolized by the first vertex encountered while descending along the time axis.'[1] This key possibility is a direct consequence of the fact that, the category of relation being geometrically schematized, stemmatic connections are rendered into *morphological* terms and positional values become therefore definable and identifiable.

1 Thom, 1980: 207.

(v) The 'capture' graph also clearly shows the proto-grammatical nature of the archetypes expressed by *elementary* actantial graphs. It corresponds to a proto-verb associated with a *canonical semantics* (in the sense of Sec. 5.3) which is not a semantics in the traditional sense but *a morphological one which generates syntax and corresponds to the internal dynamic of the generating catastrophe.*

(vi) The actantial graphs are *generic* and are realizable in space-time. *Their local morphological complexity is therefore drastically limited by space-time dimension.* This essential fact may be considered as an explanation for *the intrinsic (non-contingent) limitation of verbal valence,* which, as we noted in Sec. 4.5, is a deep linguistic phenomenon.

In order to achieve a deduction of cases, it is now necessary to *generate archetypal* actantial graphs. It is in the solution of this most central problem that the morphodynamical point of view establishes its utmost significance.

Let us recall (see Sec. 1.3) that elementary catastrophes are catastrophes whose internal dynamics derives from a potential function f on a differentiable manifold M (internal space), where the internal states are the minima of f, and where the stratified space (W, K) is the universal unfolding of a singularity of codimension ≤ 4.

Let f be a potential of finite codimension on an internal manifold M. Let (W, K) be its universal unfolding, $\Sigma \subset M \times W$ the submanifold of points (x,w) such that x is a critical point of f_w, χ the restriction to Σ of the projection $M \times W \rightarrow W$, and K the apparent contour of Σ on W relative to χ, i.e., the set of $w \in W$ such that f_w has a degenerate critical point (and is thus structurally unstable according to Morse theorem). The application $\chi : \Sigma \rightarrow W$ is an elementary catastrophe.

To a path γ in the external space (W, K), one can naturally associate an actantial graph describing the interactions between the local 'actants' which are the minima of the generating potential. For example, as shown in fig. 14, the actantial graph of capture can be generated by a path in the external space of a *cusp* catastrophe.

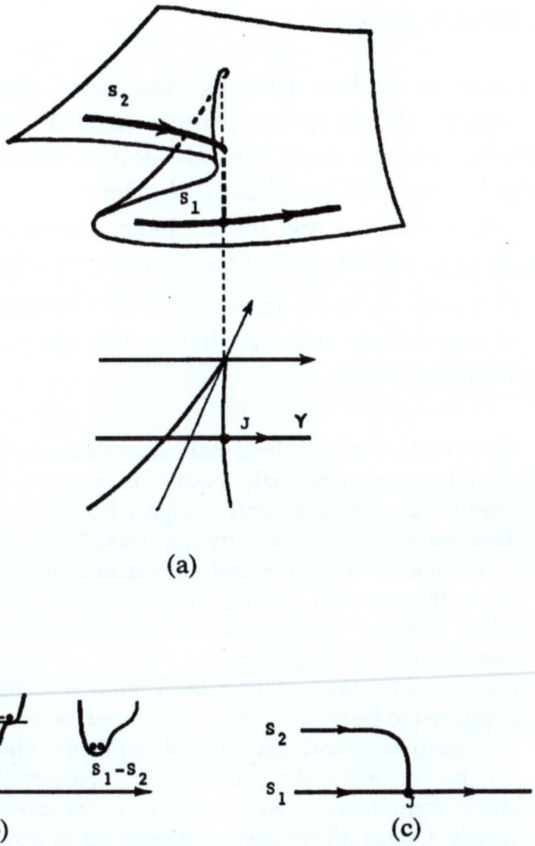

Fig. 14: Generation of the actantial graph of capture by the cusp catastrophe. (a) The path γ in the external space of the cusp. (b) Evolution of the minima. (c) The corresponding actantial graph.

Whence Thom's key idea:

By interpreting the local stable regimes as [actants], it is possible to give a semantic interpretation, expressed in ordinary language, to the qualitative aspect of the catastrophes. If the external coordinates are taken to be exclusively spatial, then the catastrophes are interpreted as substantives. If time is introduced, they are interpreted as verbs. [...] More generally, it is useful to consider the plane sections of dimension one or two in the universal unfolding of every catastrophe. We will then get what I consider to be the *universal structural table* containing all types of elementary sentences, that is to say,

> bearing an autonomous signification undecomposable into smaller units with the same property.[1]

The cogency of this idea comes from the *classification theorem* of elementary catastrophes which renders explicit the geometrical constraints imposed on the interaction between local regimes. Actually *it solves the problem of case deduction.*

Let us make a few more observations on this morphodynamical generation of archetypal syntactic morphologies.

(i) It meets up with case grammars starting from a general theory of regulation and stability in the structural realm. As Wolfgang Wildgen notes:

> The structure of the elementary interactions which are derived from paths in the bifurcation [external] space of elementary catastrophes, defines different roles which can be roughly compared to the 'schémas actantiels' proposed by Tesnière and to the 'case frames' classified by Fillmore. The basic difference between these structures and the semantic archetypes consists:
> (1) In the *preverbal* character of archetypes. The structures proposed by Tesnière, Fillmore and others are only generalizations of linguistic structures found in natural languages.
> (2) The foundations of the classification of archetypes in a formalism which is supposed to be basic for many biological systems. It is therefore universal in a very deep sense and it is of interdisciplinary relevance.
> (3) The semantic archetypes are *irreducible gestalts*. They are not composed in a single combinatorial way. This fact constitutes a major difference in Thom's theory against all theories proposed up to now. Some of these have tried to describe field-like structures, but as no tool for consequently doing so was available they all drove away irresistibly attracted by the static-logical paradigm.[2]

(ii) The difference between the actantial graphs schematizing the interactions between local proto-actants and the archetypal graphs derived from the elementary catastrophes is that in the latter, the actants are defined by the *same* global generating potential (global relative to the actants, even if it is only defined locally on the internal space). This potential warrants the *reciprocal determination* of the actants, that is to say, their structural

1 See Thom (1972a) and (1981a: 188).
2 Wildgen, 1981: 264–265.

order of coexistence and their interdependence as positional values.

(iii) Since morphodynamical models are paradigmatic categorizations, in this actantial schematism *the syntagmatic dimension converts the paradigmatic one*. The schematism, then, complies with one of the main eidetic features of structural conception, as described in Sec. 1.3.3. Conversion consists essentially, as we have seen, in introducing the time dimension (paths in external space) and in treating the local regimes as individuated entities.

(iv) Given a catastrophic model $\chi : \Sigma \rightarrow W$, there will be as many associated actantial graphs as there are *homotopy classes* of generic paths in the complementary set in W of the strata of codimension ≥ 2 of K. Syntactic events correspond to the (transversal) crossing of codimension 1 strata. There would generally exist *several* homotopy classes. In other words, these models naturally generate *systems of variants*, the *transformations* between variants being organized by singularities of codimension 2.

(v) Let (W, K) be the universal unfolding of an organizing centre f, and let us consider a section H of W not passing through f and transverse to K. The intersection $H \cap K$ glues unfoldings of singularities of codimension weaker (less singular) than f. It aggregates local models into a global one. But this global model can be generated by f, and is therefore a sub-model of a *local* model. Such a local/global dialectic is linked to the transitivity of universal unfoldings and is a major eidetic feature of morphodynamical models. It means that they constitute a formal universe where the classical oppositions simple/complex, irreducible/composite, component/system are no longer relevant. This explains why a trivalent verb like 'to give' can be an irreducible archetype and *at the same time* be composed of an emission and a reception archetype ('to give' as a causative of 'to have').

The above account shows that the morphodynamical approach can suitably mathematize the Fillmorean notion of scene in Sec. 5.2, and its localist nature. That is why, before concluding this second chapter, we would like to briefly touch upon the history and the principles of localism.

6 The localist hypothesis

The localist hypothesis (*LH*) consists in bringing under a common positional principle both the grammatical and local uses of cases. We think that it must be reevaluated.

The local/grammatical ambivalence (at least for case forms and flexions) is well known since antiquity. Applied at the deep level (and no longer only at the morphosyntactic one), the *LH* is, as we have seen, the key to case deduction. In order to sketch it, we will summarize Louis Hjelmslev's classic essay *La catégorie des cas.*[1]

6.1 The history of the LH

Case grammars go back to the Greeks. As they took the category of case (πτῶσις) to be the typical inflectional category, they endeavoured to undertake a morphosyntactic analysis of the case-markers and to determine and delimit their meaning. If we exclude the attempt to treat the adverb as a case (Chrysippus) as well as the difficulties raised by the vocative (Dionosyus Thrax), Greek cases are few: Nominative, Genitive, Dative, Accusative. The problem is to *deduce* this simple list in a principled way. As for the Greeks, cases were basically related to diathesis, they distinguished first the Nominative (ὀρθή) representing the active case (ἐνέργεια) along with three other oblique (πλάγιαι) cases representing passive cases (πάθος). They thus introduced in their classification, the fundamental axis of independence/dependence. But, in order to differentiate the three oblique cases, they needed another axis, and remained quite uncertain as regards its determination.

If the Greeks succeeded in determining satisfactorily the general meanings of the Nominative (active), the Accusative (the pre-

1 Hjelmslev, 1935.

ferred case of the passive), and the Dative (goal), they did not have the same success with the Genitive. The ambiguity of the latter led them to alternate between two methods of analysis which reappear again and again in the course of history:

(i) the method that Hjelmslev refers to as the *extralinguistic division* which introduces several underlying cases for a single surface form;

(ii) the method of *metonymy* which privileges one use of the case as its basic meaning, and considers the other meanings as derived.

Greek theory culminated in the works of Appolonius Dyscolus. It is 'indiosynchronic' to the extent that it takes into account only a single language (Greek). With Latin grammar, the situation changes, for the Roman grammarians analyzed *two* languages (Greek and Latin); the basic problem was posed by the Ablative (*casus sextus*)[1] which shares the features both of Greek Genitive and Dative. Using extralinguistic division, they ended up with additional cases. But they returned to an idiosynchronic case theory, where the properties of a single language (Latin) were supposed to be universal. Their approach weighed heavily on the grammars of succeeding generations, and produced what Jespersen called the 'squinting grammars', interpreting every idiosynchronic system from Latin. This persisted for a long time particularly in the identification of cases with flexional endings.

The *LH* was introduced by the Byzantine grammarians Theodore Gaza and Maximus Planudis who further developed the works of Appolonius Dyscolus. For Gaza, the opposition differentiating oblique cases obtains between Accusative and Genitive. Accusative means that the subject is acting in the direction of an object. It is a sort of movement outwards (ἐχπομπή). Inversely, Genitive means that the subject is acting in capturing an object. It is a sort of movement inwards (εἰσπομπή). Dative is neutral with respect to this opposition. This analysis attempted to solve two main problems:

(i) reconciling the interpretations of cases as relations to the verb and as relations between the nominal terms;

(ii) defining the cases from the perspective of the noun that undergoes case-inflection.

1 The *casus quintus* was the problematic Vocative.

It is for accounting for the second aspect that Planudis [1] introduced the most important opposition of the *LH*, namely that of distance/proximity. For him, Genitive was the distance case, Accusative the proximity case, and Dative the neutral case of 'rest'. Planudis worked out a two-dimensional classification of cases based on the two axes of independence/dependence and distance/proximity.[2]

	Proximity	Neutral	Distance
Independence		Nominative	
Dependence	Accusative	Dative	Genitive

The basic principle of the *LH* postulates that the notion of direction underlying the proximity/distance axis is *abstract*, undiscernibly local *and* grammatical, and refers equally to the concrete spatial relations between the referents of the nominal terms of a sentence as well as to their intra-sentential grammatical relations. This means that the concrete spatial relations are not the referents for cases, but rather their *schema*. The intuitive spatial notion of direction *gives form to and constrains* the category of case. The *LH*

> recognizes as equivalent the concrete or local manifestations and the abstract or grammatical manifestations of the principle of direction. (p. 15)[3]

According to Hjelmslev, the main contribution of the Middle Ages was to recognized that 'each flectional element is simultaneously syntagmatic and paradigmatic' (p. 22), specially in noticing, on the basis of syntagmatic facts, that Nominative is also governed by the other terms of the syntagm and thus belongs, like the oblique cases, to the category of dependence: every case can be either governed or absolute, dependent or independent.

Similar reflections were further deepened during the 19th century. Bernhardi discovered the relationship between cases and *prepositions*. Roth reformulated the dependence/independence opposition. Ast and Döleke went back to the *LH*: the principle of

---

1 The Greek text of Maximus Planude's grammatical works can be found in Bachmann, 1828.

2 Planude's terminology is rather confusing since proximity is ἐχπομπήand distance εἰσπομπή.

3 In this section the pages of Hjelmslev, 1935 will be referred to in the text.

direction is manifested simultaneously at the concrete (local) level and at the abstract (syntagmatic) one. Hermann attempted to deduce cases from the Kantian categories. Wundt separated the problem of cases from that of their morphosyntactic expression by case-markers in noting that they can be expressed also by word-order.

However, the major contributions came from the Indo-European linguists and the panchronic theorists (e.g., Humboldt) who broke away from the 'squinting grammars' by showing that the Greco-Latin theories involved an analysis of only *particular* idiosynchronic states. According to Hjelmslev, the *LH* achieved a sort of perfection with Bopp and Wüllner.

Bopp wished to show that case-markers historically developed from an agglutination of pronominal roots, which were themselves derived from local adverbs or prepositions. He regarded cases as

> primitive expressions of spatial order, capable of conforming to the requirements of a more developed thought operating in the sphere of time and causality. (p. 36)

Wüllner, influenced by Kant's Critique gave a deep epistemological form to this viewpoint:

(i) The linguistic phenomenon is *subjective* in the *transcendental* sense: the speaking subject (which is not a psychological one) selects the grammatical forms according to a principle imposed by its 'conception' of objective reality.

(ii) The conception underlying a grammatical form must be a unique and sufficiently abstract idea from which every concrete use of the form can be deduced.

(iii) As far as the case form is concerned, the underlying conception is the *spatial* conception applicable to spatio-temporal phenomena as well as to syntagmatic rection.

Contrary to this localist point of view, the anti-localist theories assumed a purely *categorial* conception of grammar. At Hjelmslev's time (1935) the dominant theory was in fact the syntactic theory (Theodore Rumpel) which projected the logical forms of judgement into linguistic immanence.

6.2 *The* LH *and the spatial conception*

Hjelmslev virulently criticized the categorial conception of language. Of course there exists 'an intimate relationship between the categories of language and the categories of thought' (p. 29). But this is not however an identity.

> All that we can say is that language is in the last analysis an epistemological system and that consequently, the deepest concepts of language are basically of the same nature as the ultimate concepts of logical analysis. (p. 29)

Hjelmslev lays heavy emphasis on the necessity of identifying the linguistic categories by an *empirical* method avoiding a mere superimposition of the categories of thought on it.

> It is useless to assume that linguistic description can avoid all epistemological notions . The form of language is a categorial form. But this does not mean that this categorial form is anterior to language. It means on the contrary, that there exists a specific categorial order which is manifested by language, and only by it. [...] It follows that there are two possible methods of establishing tables of categories: firstly, there is an *exclusively linguistic* method which has the limited scope of determining, by an empirical procedure, the categories reflected directly in language, i.e., the categories of notions which correspond exactly to the categories of linguistic expression; and secondly, there is a method which is *both linguistic and extralinguistic,* which, using artifices, attempt to establish categories which are not directly present in language, which can be expressed linguistically only by a circumlocution. From the linguistic point of view, the first can be called *immanent,* and the second *transcendental.* The transcendental method proceeds according to the principle of extralinguistic division. (p. 49–50)

This criticism is coupled in Hjelmslev with a criticism of comparative grammar which adds to the extralinguistic division, the diachronic division of evolutionary chance, and which 'has the special feature of not being a grammar' (p. 59). Hjelmslev aimed at a strictly structural analysis of grammar. For him, every state of language constitutes an idiosynchronic system whose units should have an empirically determinable meaning. In what concerns the category, we should not try to unify the multiple uses manifested by case markers which vary considerably cross-linguistically. We must first of all isolate *the global meaning* of the category as such, and then – as in phonology – determine the way in which each idiosynchronic state of language partitions this semantic space into cases. It is the category that is universal and not any one of the particular case systems.

> It is generally impossible to maintain that each language possesses some cases whose meaning would be delimited once and for all. To establish a general grammar of cases, it is enough to fix the fundamental meaning of the category taken as a whole, and then see how the cases belonging to different languages behave in relation to this range of signification. There are no universal cases. It is the category that is universal. (p. 69–70)

The case problem must be disconnected from that of expression (prefixes, inflections, prepositions, word order) and we must treat the case data (as any other linguistic data) as a fact of *value*. This is the basic structural principle: 'a linguistic form is an *expressed value,* [...] a grammatical category is defined by value, not by expression (p. 77)'.

Applied to the category of case, this principle says that the value of a case 'is not identical to the sum of its uses at the level of expression' (p. 88). The value of a case is the minimal difference of meaning which renders possible its use.

Such a perspective leads us to a purely *semantic* definition of cases. Its goal is to set up a *fundamental* case system underlying the particular systems. The deep meaning of the category of case is not a psychological fact. Though its form is subjective (in the Kantian sense), the system is nonetheless 'objective'. It is an 'abstract and virtual reality' (p. 88) which, according to Hjelmslev must be explained with the help of 'Platonic ideas' (p. 86).

It is for determining the fundamental meaning of the category of case that Hjelmslev took recourse to a strict form of the *LH*. 'Case is a category which expresses a relation between two objects' (p. 96), the category of relation being here founded on the primitive and abstract notion of *direction*. The logical or grammatical cases express 'an immediate relation or a negative behavior' with regard to the 'intimacy' of the connection linking the objects, while the topical or local cases express 'a mediated relation or a positive behavior' (p. 97).

Hjelmslev reached this conclusion empirically by analyzing a large number of case systems, but also theoretically for general epistemological reasons: projecting the predicative structure of judgement into grammatical immanence is a dramatic mistake, *for ultimately it is the logical order that is extralinguistic and not the spatial conception*. His main observation is that spatial intuition *must* constrain the abstract category of relation:

> Spatial conception is inevitable if one wishes to give the relation in abstracto a tangible and plastic interpretation. To stick to abstract relations without providing them with a substratum where they can be represented, is to prohibit in advance the clear and evident explanation of facts. (p. 45)

To summarize we can say that Hjelmslev considered the *LH* as a synthesis between:

(i) a structural approach to grammar;

(ii) an imperative of spatial schematization.

He applied to the case-primitive of 'direction' the structural theory of oppositions as per which the differential features are not logical oppositions, but rather oppositions between the neutral and complex terms of 'participative' oppositions *A*/(*A* and ~*A*). He showed with the help of very concrete examples (particularly from Eastern Caucasian languages, Tabassaran and Lak) how particular case systems are specifically organized, and how the localist structural approach can explain the observed syncretisms in an intra-systematic way.

6.3 *The paralogism of* La catégorie des cas

Epistemologically, the *LH* subordinates structural syntax to a *schematic* spatial conception. Such a conditioning is possible only by means of a third term, which like every schema, should be on the one hand, akin to conceptual categories, and on the other hand, akin to perceptive intuitions. It should also, in order to match a figurative case relations with semantic roles, be able to develop the *geometry* of positional relations. Without the *LH*, one fails to establish a *principled* case deduction. But without a geometry of possible interactions between positional proto-actants, one fails to found the *LH*. Structural syntax as a whole points – as if towards its genetic principle – towards a geometry of position.

In this sense, Hjelmslev's interpretation of the *LH* is quite *paralogical*. On the one hand, it asserts that spatial intuition gives syntax its form, but on the other hand, it immediately negates this assertion by defining *semiotically* the meaning of the case category.

Hjelmslev brings back therefore to a semiotics of space – and not to a geometry – the foundational spatial conception.

Contrary to Hjelmslev, we think that structural syntax should resolutely opt for a transcendental method proceeding according to the principle of extra-linguistic division. It has to geometrize the *LH*, and we have already seen that it is possible using *CT*.

CHAPTER III

Semio-narrative structures

1 Introduction

In the previous chapter, we investigated the eidetic and formal content of the notion of double articulation (phonological and syntactic). We concluded that in both cases, for reasons as much practical as theoretical, methodological, and epistemological, it was difficult to avoid a morphodynamical conception.

This double application of the morphodynamical schematism in the double articulation of language proceeds from a double interpretation of the catastrophist schemas. The phonological second articulation is a particular (and complex) case of a control situation σ : $W \rightarrow \mathcal{X}$, where an external space W controls internal dynamics X_w whose attractors define the internal states of a perceptual 'black box'. When W is a pure space of control, we get models of critical phenomena (like phase-transitions in thermodynamics and caustics in optics). When W is the spatio-temporal extension of a material substratum and when the internal dynamics define the local 'metabolic' regimes, we get models of morphogenesis (for instance, embryogenesis). When W is simultaneously a control and a substratum we get either behavioral models (for instance Zeeman's ethological models), or categorization models (as for categorical perception). On the other hand, when W is a purely ideal space and when the local regimes are interpreted as abstract identities, the catastrophic schemas become generative devices of actantial relations, and therefore of the first articulation of language. In this context, *we will use the term 'conversion' for this possibility of transforming taxonomic models of categorization into actantial schemas*.

In this last chapter we shall extend our enquiry on the relevance of a morphodynamical schematization of structural categories to Greimas' theory of semio-narrative structures.[1]

As we already indicated in Sec. I.2.7, Greimasian theory of narratives is structural-generative and syntactico-semantic. According to the general principle of a hierarchy of levels of organization, it distinguishes on the one hand, the deep syntactico-semantic [2] level of semio-narrative structures from, on the other hand, the surface level of discursive structures where appear the most prominent features of narrativity, such as actors, thematic investments, intrigues, the spatio-temporal distribution of events, forms of figurativization, the procedures for creating effects of 'reality', etc. The deep syntactico-semantic level – which is theoretically the most important – further consists of two levels: the level of taxonomic and syntactic (logico-combinatorial) organization of deep contents (the semiotic square) and the level of deep actantiality (what Greimas called 'anthropomorphic' syntax). They are related by the conversion of deep logical relations into actantial ones (see, Sec. 2).

Greimasian theory is an achieved example of a structuralist construction, in the strong Hjelmslevian sense. It is based on certain primitive concepts (considered as undefinable) which are categories of structuralism (relation, difference, junction, position, etc.) that have to be schematized.

As regards the actantial level, Greimas' conception is case-based (in the sense of Tesnière and Fillmore), but in a very abstract fashion. This is due to the nature of its object. Compared to linguistic actantial structures, where the case analysis should account for the extreme heterogeneity of verbal lexicon, narrative actantial structures are deeper and more universal. They are restricted to a very limited number of archetypal interactions between positional proto-actants (conjunction/disjunction between subject and object, conflict between subject and anti-subject, transfer and exchange between sender and receiver, etc.). In other words, narrative actantial universals are of

---

1 These reflections first appeared in 1977 under the title 'Topologie du carré sémiotique' in a volume of *Etudes littéraires* (Quebec: University of Laval) edited by Pierre Ouellet (see, Petitot, 1977b).

2 The concepts of semantics and syntax share a rather peculiar use in structural semiotics, different from their classical linguistic and logical one. It will be progressively defined below.

such a nature that the question of their *closure* as well as their *deduction* is well posed, the proliferation of verbal lexicon being taken back to the discursive component. They largely coincide with what we referred to as the *canonical semantics* of case-universals (Sec. II.3.3).

As regards the (taxonomic) deep level of fundamental semantics, the logico-combinatorial articulation of contents is understood, following Hjelmslev, in paradigmatic terms, as in phonology. But contrary to what happens in phonology, where as we saw, form is identifiable with the organization of substance, and is, at least partly, amenable to experience, the *form of meaning* is articulating a substance which *cannot* be empirically observed. The form of meaning cannot be objectivized in a classical manner. Its hidden immanence has to be identified with a metalinguisitc reconstruction, and it is therefore particularly necessary to *schematize* the categories (the primitive undefinable concepts) on which this latter depends.

Thus, the case of semio-narrative structures differs doubly from what we previously developed in Chap. II.

(i) The articulation of contents (semantic categorization) can no longer refer – in contrast with phonology – to models sharing an empirical value, but only to a schematization. In this framework, Hjelmslev's parallelism between the expression and content planes has to be reinterpreted as a principle according to which the semantic schematization must depend on the *same eidetic* as phonological modeling.

(ii) The actantial narrative grammar is a *pure* structural syntax in which – in contrast with case grammars – the diversity and heterogeneity of the verbal lexicon are reduced from the outset.

Moreover, as we strongly emphasized in Sec. I.3.3, in the semio-narrative realm, the relation between syntax and semantics is thought of as a projection – a conversion – of the synchronic paradigmatic axis on the diachronic syntagmatic axis, the paradigmatic infrastructure being a semantic stance that:

(i) defines the 'deep meaning' and the anthropological function of narratives, and

(ii) constrains the syntagmatic unfoldings.

This structuralist axiom prescribes the formalization of narrative syntax to be founded on the *same* sort of mathematics as paradigmatic categorizations, and this is a challenge difficult to tackle.

These considerations justify the following principles and hypotheses.

Hypothesis H1: *Principle of schematization of the undefinables.*
The semio-narrative theory has to schematize its undefinables using mathematical structures akin to paradigmatic categorizations and actantial syntax.

Hypothesis H2: *Hypothesis of a morphodynamical schematization.*
The categoriality of semio-narrative theory being structural, relational and positional, its schematization will depend on a 'geometry of position' and morphodynamical modeling.

Hypothesis H3: *Principle of mathematization of non-primitive concepts.*
From the schematization of the undefinables, derived concepts can be mathematized in a non trivial way.

Hypothesis H4: *Principle of conversion.*
In the framework of a morphodynamical schematization, the conversion of the 'logical' operations constituting the semiotic square into an 'anthropomorphic' syntax should be derived from a *double* interpretation – categorical and actantial – of the *same* catastrophic schemas.

Let us clarify better the problem of mathematizing theoretical concepts, problem which is largely misunderstood in the social sciences. Let us go back to the parallel with Physics. There was a time when physical theory was a conceptual descriptive theory derived from primitive concepts. This 'physics' appears to us today as a metaphysics of nature. Modern physics began when the primitive concepts and the principles of experience (causality, invariance, symmetry, relativity, etc.) could be adequately mathematized. It required new appropriate tools, and in particular, integral and differential calculus. It destroyed the conceptual simplicity and the metaphysical organization of nature. It introduced a shift from a closed conceptual theory to an unending open algorithmic theory which constantly addressed new problems. In particular, after the brilliant success of Newton, it encountered the non-integrability of the *n*-bodies problem which gave rise to the extraordinary developments of qualitative dynamics since Poincaré.

This is enough to show that the function of simplicity in a theory which proceeds by mathematizing concepts is not at all of the same order as in a conceptual-descriptive theory. It bears on the mathematization of undefinable primitives and principles, and often leads to great difficulties in the process of its complexification.

We think that a mutation of the same order is henceforth possible for structural semiotics. The deep interaction between the empirical, mathematical, and epistemological levels which constitutes the pillar of objective determination in classical sciences can be shifted to the field of narrative semiotics once we have at our disposal a mathematical eidetics which would be to structuralism what integro-differential calculus is to classical physics. Of course, one could think that in semio-linguistic matters, such a formalization already exists and consists in symbolizing the undefinable concepts logically. But this is a mistaken prejudice which embeds structural theory into a formal universe where the combinatorial complexity is free, while in structural organizations complexification is submitted to very strong constraints.

We cannot underestimate the importance and the general relevance of various formal explanations of complexity. We have to investigate their validity for the structural realm. Our hypothesis is that the logico-combinatorial explanation is almost empty. Of course narratives can be conceived of as assemblages of sequences and this combinatorial complexity can be analyzed. But the main theoretical problem concerns the nature and the deduction of the elementary semio-narrative structures, and at this level, logical symbolization is not relevant, for it is trivial. Instead, the morphodynamical schematization explains what is to be elementary in a non-trivial way.

We will deal with the technical problems of modeling in another book.# In this chapter we will show how the conceptual organization of Greimas' theory appeals to the hypotheses *H1–H4* described above. For an introduction to this theory, in addition to the celebrated Greimas-Courtès dictionary (1979), the following works could be

Petitot, 1992.

consulted: Greimas, 1970, 1976a, 1976b, 1983, Courtès, 1976, Coquet, 1982, Hénault, 1979, 1983.

2 The sources of Greimasian theory

As we already mentioned it in Sec. I.2.7, Greimasian theory acknowledges two sources: the works of Vladimir Propp on Russian folktales and those of Lévi-Strauss on myths. From the former, Greimas adopted the idea of a syntactic organization of narratives, and from the latter, the idea of a semantic component. In this way, as suggested by J. Courtès it is 'situated at the meeting point of anthropology and linguistics'[1] and aims at describing at a deep level the syntactico-semantic *form* of semio-narrative structures.

From his analysis of a large corpus, Propp [2] had concluded that folktales could be described as chains of typical sequences whose invariant elements are referred to as *functions*. He notes

> the constant and permanent elements of the folktale are the functions of dramatis personae, whatever they are, and in whatever manner these functions are fulfilled.[3]

By submitting Propp's method to a rigorous critical evaluation, Greimas reached the conclusion that the Proppian functions were not sufficiently well-defined. They are at times actions (e.g., hero's departure) and at other times states (e.g., lack).[4] Hence the necessity of a further elaboration. Following the standard strategy of distinguishing between the deep and surface structures, he assumed that

1 Courtès, 1976: 28.
2 See, Propp, 1928.
3 See Chabrol, 1973: 14.
4 See, Greimas, 1976b.

> this elaboration appears as a syntactico-semantic representation, simultaneously encatalyzed and disambiguated, operating as a *deep structure* relative to the surface structures that are the textual occurrences (tokens).[1]

The subsequent normalization of the Proppian functions resulted in an actantial component generating narrative statements whose syntagmatic concatenation in turn generates the characteristic sequences of the tale. The syntactic organization deduced from Propp constitutes the level of the 'anthropomorphic' syntax, an actantial and event-based syntax which is a 'surface' one relative to the deep semantic level of the semiotic square, but a 'deep' one relative to the discursive level of the actors (the dramatis personae) and their thematic roles, procedures of figurativisation, and discursive configurations. Propp's and Greimas' merit is in having successfully separated a discursive component responsible for the rich diversity of tales from a syntactic, much more abstract and rigid, one that can be subjected to a rigorous structural analysis.

The 'anthropomorphic' syntax raises delicate problems analogous to those posed by case-grammars: topological definition of positional proto-actants, configurational definition of their relations, geometrical schematization as opposed to graphical symbolization, deduction of actantial structures (Proppian functions), etc. However, at the *global* level of what Greimas referred to as the *narrative schema*, there are new and original problems. Indeed, the syntagmatic organization of tales is *finalized*. Things happen as if they were drived by an 'intentionality' starting from an initial unstable state of lack to a final stable state where the lack is liquidated and the equilibrium restored, along an invariant canonical path. It is by means of such a schema that the human 'imaginary' stance represents in narrative terms the 'meaning of life' as 'quest' (successful or failed).[2]

It is in the understanding of this finalized global structure that Lévi-Strauss contributed decisively while showing that there existed in myths *globally determinant paradigmatic operations driving the syntagmatic sequentiality*. In order to include the latter in the theory, Greimas projected into the deep level these paradigmatic articulations of contents (semic categories and semiotic square), and posited

---

1 Ibid.: 8. All the concepts and problems considered in this section are summarized in Sec. 3.

2 Ibid.: 9–11.

logical abstract operations under the anthropomorphic doings. Already reduced to its actantial skeleton by the transfer of surface contents to the discursive component, the narrative organization now appears as a syntagmatic unfolding of deep logico-semantic operations of a paradigmatic and global nature. The deep semes are of a nature entirely different from those familiar in componential lexical analysis. As the raw and the cooked of the celebrated Bororo myth analyzed by Lévi-Strauss, they are metalinguistic, categorical, and classifying semes, pertaining to what Lévi-Strauss called *codes*. In a way, they constitute the 'unconscious' ground of the tale. They regulate the global syntagmatic units. It is in this respect that they allow us to say that the narrative structures constitute an intermediate 'autonomous instance' between

> the *ab quo* fundamental instances, where the semantic substance receives its first articulations, and the *ad quem* instances, where the signification is manifested across different languages.[1]

The projection of the paradigmatic axis onto the syntagmatic one and the categorical treatment of deep semantics have often been criticized. One of the most relevant evaluation is due to Thomas Pavel. Pavel observed [2] that, contrary to what is the case in phonology, if we consider mythic sequences as 'allophones' of mythical 'phonemes' ('mythemes') we get a 'phonology' *without an underlying 'phonetics'*.[3] He pointed out that generally speaking, every theory involving a movement from an etic to an emic level [4] must – for being consistent – satisfy a number of conditions, and particularly the following:

> the rules establishing the correspondence between the etic and emic levels should be able to associate each etic description with a single emic description and, excluding cases of free variations, *vice versa*.[5]

Now, according to Pavel, this condition of reciprocal translatability is *not* satisfied in Lévi-Strauss' analysis.

---

1 Greimas, 1970: 159.
2 See, Pavel, 1979.
3 See, Sec. 2.1 for a discussion on this problem.
4 For the etic/emic distinction, see Sec. II.2.1.
5 Pavel,1979: 663.

It is precisely this very delicate problem that Greimas' theory of semio-narrative structures undertakes to solve. For this, it had, first of all, to deepen the actantial component and reduce the verbal semantism to actantial relations which are of *the same order of abstraction* as the immanent semic articulations. Indeed, Pavel's criticism is valid only if the narrative events are thought of as heterogeneous owing to the differences in their verbal content. But if we assume that these events are the surface (discursive) manifestations of deep actantial invariants then the criticism fails. It is therefore important to separate the actantial level from the discursive component and take as elements of the narrative syntactic component, only actantial relations which are both primitive and abstract (conjunction, disjunction, transfer, etc.). Once we have reduced that way the verbal semantism, we can return to Lévi-Strauss' approach and assume that the syntagmatic concatenations of sequences narrativize 'logical' operations on paradigmatically interdefined deep contents. As Greimas and J. Courtès acknowledged, Lévi-Strauss' celebrated interpretation of the Oedipus myth [1]

> allowed us to account for the existence, at the depths of discourse, of semiotic structures containing a fundamental semantics and syntax; at the same time it made the mythic discourse lose its specificity: for similar semiotic structures govern poetic, oneiric, etc., discourses. Consequently, the practical/mythical dichotomy loses its relevance; the practical level is identified with the figurative dimension of discourse, while the mythic level corresponds, in the generative pathway, to the deep semiotic organizations.[2]

In *Du Sens*, Greimas commented in great detail on Lévi-Strauss' analysis of the reference Bororo myth of the bird-hunter in *The Raw and the Cooked*, the first volume of the *'Mythologiques'*.[3] He showed how the syntagmatic algorithm is globally controlled by an operation of *inversion* of topical contents. The two inverted contents are correlated with two other contents (whose transformation is similar to that of the topical contents) characterizing the initial and final sequences of the story (see, Table 2).

---

1 Lévi-Strauss, 1958.

2 Greimas-Courtès, 1979: 241, item 'Mythique'.

3 For the links between Lévi-Strauss' *Mythologiques* and Greimasian theory, see also Courtès, 1973.

Contents	Inverted contents			Posited contents		
	Correlated content	Topical content		Topical content		Correlated content
Narrative sequences	Initial sequence	Nest of spirits	Nest of macaws	Return	Revenge	Final sequence

Table 2: From Greimas, 1970: 198.

This example shows the distribution of the levels of organization of the tale. At the paradigmatic level of deep contents appear Lévi-Strauss' *codes*.

> The code is a formal structure (1) constituted of a small number of semic categories and (2) whose combinatorics is able, by generating sememes, of accounting for the invested contents taking part in the dimension selected by the mythological universe.[1]

In the Bororo reference myth, the code for alimentary categories is, for instance, given by the hierarchy of oppositions:

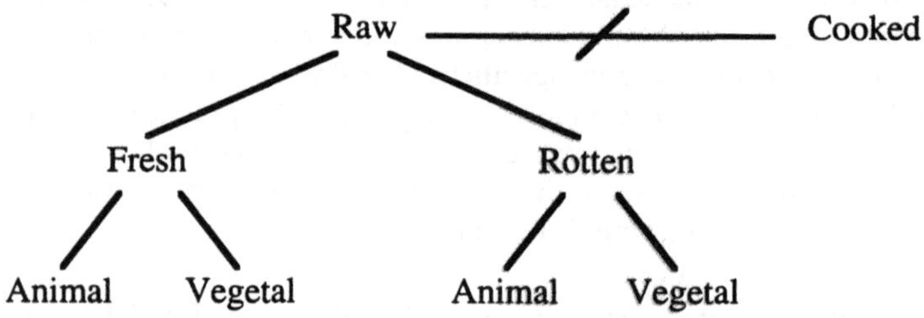

It generates sememes like 'Raw + Fresh + Animal', 'Raw + Fresh + Vegetal' etc. Now, as Lévi-Strauss has shown, these sememes are assumed by lexemes (animal actors) according to the following correspondence:

'Raw + Fresh + Animal'→Jaguar,
'Raw + Fresh + Vegetal'→Stag,
'Raw + Rotten + Animal'→Vulture,
'Raw + Rotten + Vegetal'→Tortoise.

1 Greimas, 1970: 196.

These lexematized actors share a *double* function: on the one hand, as bearers of sememes derived from the code and on the other hand, as actants.

By achieving some of Propp's and Lévi-Strauss' proposals within the framework of structural methodology, Greimasian theory opened a new rational horizon for narrative semiotics. It enabled us to conceive of narratives as a 'window' on the anthropological structures of the imaginary stance.

> The narrative considered as the irruption of a discontinuous form into the discursive permanence of a life, a history, an individual, a culture, segments it into discrete states between which it situates transformations: this permits us to describe it, in the form of statements of doing affecting the statements of state, the latter guaranteeing the semiotic existence of the subjects in junction with objects invested with values.[1]

Or, put it another way:

> Actantial narrativity appears more and more like being able to account for the organization of human imaginary stance, a projection of collective as well as individual universes.[2]

That is why, it can be said that narrative competence is a *cognitive* one, manifesting what Greimas referred to as a 'syntagmatic intelligence'.

3 Panorama of Greimasian theory

In this section we will briefly indicate the main aspects of Greimasian theory. This will lead us to a list of technical problems that we will attempt to solve in another book.[#]

1 Greimas, 1973a: 34.
2 Greimas, 1973b: 162.
Petitot, 1992.

3.1 *The generative pathway*

Like most of the theories of language, the Greimasian theory defines, analyzes and links up several hierarchized levels of organization. It identifies, as we have seen, basically three levels, each sharing a syntactic and a semantic component. The deepest level is the logico-semantic one, concerned with the semic categories articulated by the semiotic square. The next level involves the conversion of logical operations into syntactic doings. It is the actantial 'anthropomorphic' level of narrative syntax. These two levels constitute the immanent level of the semio-narrative structures. As for the discursive 'surface' level, which proceeds from the semio-narrative structures to the manifestation, it is divided into

(i) a discursive syntax defining the procedures of actorialization of the actants, the temporalization and spatialization of the narrative events, and

(ii) a discursive semantics defining procedures of thematization and figurativization. The global organization of the theory, referred to by Greimas as the *generative pathway*,[#] can be represented as in table 3.

The French expression 'parcours génératif' will be translated by 'generative pathway'. The idea is that of a hierarchized series of levels going from deep ones to surface ones.

		Syntactic Component	*Semantic Component*
Semio-Narrative structures	Deep level	Fundamental syntax	Fundamental semantics
	Surface level	Surface narrative syntax	Narrative semantics
Discursive structures		Discursive syntax: Actorialization Temporalization Spatialization	Discursive semantics: Thematisation Figurativization

Table 3. From Greimas and Courtès, 1979: 160.

3.2 *Fundamental semantics*

Along with the fundamental syntax (semiotic square), the fundamental semantics constitutes the instance *ab quo* of the generative pathway. It is an inventory of semic categories which can be actualized at the narrative level.[1] Its components are *semes* (units of content) defined in a *relational* way by their differences.

3.2.1 *The notion of seme*

Following Hjelmslev's principle of parallelism between the expression and content planes, Greimas conceived of the content plane as a substance articulated by a categorical form. The problem is that the substance of content is not an undifferentiated matter (*hyle*). It is highly organized and the structuralist principle of the primacy of form over substance asserts only that the semiotic *values* are paradigmatically defined by categorizations. But as was the case in phonology, we can suppose that these categorizations are not reduced to a pure relational form plastered onto an amorphous substance, but on the contrary, *emerge* from the self-organization of the latter.

1 See, Greimas-Courtès, 1979: 330.

The theoretical clarification of the notion of seme presupposes therefore the notion of categorical articulation that allows semantic continua to become discrete through emerging discontinuities. But, as J. Courtès observes, 'discontinuity is particularly problematic in the domain of semantics'[1] because 'from the very beginning, all 'objective' reference to the signifier is excluded'.[2]

Greimas laid much stress on this point, and relying on the principle of parallelism, treated discontinuity as an *undefinable* that 'must be retained in the epistemological inventory of non-analyzed postulates.'[3] Hence a first problem : How to conceive of articulation, categorization, discretization of the substance of content as an immanent operation of discontinuity prior to manifestation and independent of expression?

A purely conceptual-descriptive theory cannot furnish an answer to this problem. Certainly, one would insist, as in the case of phonological distinctive features, that the nature and function of the semes are purely formal, distinctive, and relational, that the 'seme is not an atomic and autonomous element [and] owes its existence entirely to that which differentiates it from other semes',[4] that it is only a position in a network, and that the semic categories contrasting two semes are 'logically anterior' to the semes that they constitute. But, if lacking any objectivization of their relational form, one can define them only by a metalinguistic naming, and this opens up a vicious circle. Hence a second problem : How to conceive of relational forms which actually precede the contents on which they operate and the terms that are defined by their operation?

3.2.2 *Semic categories*

The structural primacy of relations over terms leads us to reject any atomic and autonomous existence of the semes, and to define them as terms of semic categories. In a semic category s_1/s_2 (also called elementary structure) the semes s_1 and s_2 are defined by their relation, which is simultaneously a conjunction and a disjunction, s_1 and s_2 being the opposite poles of the semantic axis that links them: the

1 Courtès, 1976: 45.
2 Ibid.: 46.
3 Greimas, 1966:18. See Sec. 5.2.
4 Greimas-Courtès, 1979: 332.

semic difference is a relation of reciprocal presupposition. Hence a third problem : How to conceive of relations of junction (conjunction/disjunction) and of reciprocal presupposition which are 'pure', that is to say independent of the specific semes whose value they determine?

3.2.3 *Nuclear semes, classemes, sememes, lexemes and isotopies*

Besides the delicate theoretical problems posed by its relational conception, the notion of seme displays another ambivalence. The semic dimension is located at the meeting point of immanent codes underlying discourse and lexical semantics. Every tale narrativizes the transformations and operations undergone by semes which are not manifested components of its lexemes and figures. Then, we should understand how it is possible to decompose a lexeme into semic sub-units in a manner radically different from classical componential analysis.

The answer to this problem is provided by the theory of the *sememe* conceived of as the combination of a nuclear semic figure and a classematic base. 'Resulting from the historical development of a natural language',[1] a lexeme is, in general, a fusion of several sememes which are units of the content plane. The nucleus of a sememe is a semic figure that syntactically organizes a certain number of semes, referred to as *nuclear semes*, which are *figurative* ('exteroceptive'), and which constitute what Greimas called the *semiological* level.

> They correspond to the elements of the expression plane of the semiotics of the natural world, that is to say, to the articulations of the sensory orders, to the sensible qualities of the world.[2]

The remaining part of the sememe is *contextual*. It is a classematic base constituted of contextual semes associated with 'effects of meaning' (connotations) exceeding the nuclear figure. These contextual semes are called *classemes*. They are produced by a semic 'resonance' between at least two lexemes and are associated with equivalence classes of contexts. The definition of a sememe as the

1 Ibid.: 335.
2 Ibid.: 333.

combination of a semic nuclear figure and a classematic base implies that

> the sememe is not a unit of signification delimited by the dimensions of the minimal sign; in immanence, [...] it is only a semic figure: it is only at the moment of its manifestation in the discourse that this figure joins up with its classematic base [...] and thus selects a *sememic pathway* which actualizes it as sememe.[1]

Classemes are abstract (interoceptive) semes constituting what Greimas called the *semantic* level (as opposed to the semiological level of the nuclear semes). They are units of content, which do not

> refer to any exteriority, but which on the contrary, are used to categorize the world and to supply a signification for it.[2]

They are, in the strictly relational sense of the term, contextual, interoceptive and categorizing semes, and contrary to the figures of the natural world they operate at the deep level. In other words, the semes which are *delocalized* in the syntagmatic chain and which globally control its 'intentionality' are the *classemes*. As their name suggests, the figurative semes are present only at the level of the discursive component and are therefore basically *irrelevant* for semio-narrative structures.

This conception of a deep semantic level constituted of 'free' classemes ('free' in the sense they are no longer bound to particular lexemes) that are syntagmatically delocalized and exerting a *global* function of narrative regulation, raises one of the most central difficulties of Greimasian theory. The *recurrence* of classemes defines the *semantic isotopy* of the discourse, which is very different from the semiological isotopies (which are much more evident and intuitive) obtained by iterating nuclear semes. A semantic isotopy is fundamentally *distributive*: it drives the syntagmatic distribution of figures. This can be understood only if like Greimas,

---

1 Ibid.: 335.

2 Ibid.: 333–334.

(i) we treat the 'free' classemes as *simple* semes;

(ii) we organize them using the fundamental syntax of the semiotic square;

(iii) we convert the logical operations of this syntax into actantial sequences;

(iv) we *feed back* the deep semantic (narrative) level onto the surface semiological (discursive-figurative) level *by identifying the initial simple 'free' semes with the classemes recurrently disseminated in the lexemes, that is to say, with 'bound' classemes.*

The difficulty raised by such a feed-back 'loop' of the deepest semantic level on the most superficial semiological one, has been well formulated by Per Aage Brandt.

> The distinction between the semic and lexematic dimensions [...] goes back to an enigmatic distinction in *Structural semantics*, between the 'global semantic level' and the 'global semiological level'. [...] The application of a selecting classeme to a lexeme produces the sememe; the application of the semantic level to the semiological one by means of narrative grammar produces the discourse. But in order for this process to take place, it is necessary that the selecting classeme could be *recognized* in a classeme of the lexeme; it is therefore necessary that the seme belonging to the 'semantic' level coincide with the seme that is lexematically inscribed at the 'semiological' level.[1]

In our opinion, this vicious circle is due to an ambiguity between the structural Lévi-Straussian conception of semantic codes and the classical componential lexical analysis. Indeed, despite their correlation, these two dimensions of semantics are radically different. The former involves 'unconscious' codes that globally regulate the narratives and belongs to an anthropology of the imaginary stance. The latter involves, instead, the sensorial and cognitive organization of the natural world and is part of a psycho-semantics. It would be therefore desirable to modify the terminology. As for lexical semantics, we might go back to Bernard Pottier's nomenclature, as per which the nuclear semes (figurative and exteroceptive) are *sememes* and the 'bound' contextual semes classemes. The combination of a nuclear figure composed of semantemes and a classematic base

---

1 Brandt, 1976: 148–49.

composed of 'bound' classemes produces a sememe.[1] We could thus reserve the term 'seme' for the selecting, categorizing and interoceptive 'free' classemes constitutive of the semantic level and to their treatment as minimal units by the fundamental syntax. The problem with deep 'free' classemes is that, according to Greimas, they are reducible to a pure disembodied categorical form without substance: like all formalisms, Greimas' structuralism is an idealism. Now they don't lack substance. On the contrary, only their substance can explain their anthropological function.

To clarify this point, we can use René Thom's notion of *'prégnance'*.[#] In ethology, one says that a form is 'prégnante' if its recognition is vital for survival and sets off reactions with great amplitude. For an animal, the 'prégnantes' forms are innate and are basically limited to predators, preys, and sexual partners. Besides these forms endowed with a vital signification, there is also an open and diversified universe of perceptually *salient* forms characterized by their morphology. Now, Pavlov's experiments show that a biological 'prégnance' (for instance, food) can be transferred on to a salient stimulus (the ringing of a bell, for instance), making the latter a secondary source of 'prégnance'. Thom has further proposed that a 'prégnance' is to be conceived of as a sort of fluid spreading by similarity and contact (i.e., by metaphor and metonymy) in the universe of perceptually salient forms. According to this view, the investment of a perceptual stimulus by a 'prégnance' deeply transforms its meaning and its value for the subject concerned.

In animal ethology, biological 'prégnances' are confined to typical and innate forms. It is not the case for man, with which anthropologic symbolic features relay 'prégnances' :

(i) language learning entails a 'generalized catastrophe' of biological 'prégnances', making their sources 'unconscious' drives;

1 Greimas-Courtès, 1979: 334.

In English, the Latin term 'prægnans' in the sense of 'fruit-bearing' or 'full of meaning' has been essentially restricted to female fecundity. In French and in German it kept an abstract general use. In psychology it means the force of a structure which forces itself upon the mind. These meanings exist also in erudite English. The Webster gives: 'significant quality' (meaningfulness), 'latent potentiality'. We will keep the French term.

(ii) being no longer programmed to be invested in particular forms (contrary to animals, man is 'without instincts'), biological 'prégnances' spread in the phenomenal field, and their paths, displacements, confinements or wanderings individuate the subject.

(iii) As for the subject, the investment of objects with 'prégnances' transforms them into objects of desire (of quest).

Though too rudimentary, these few remarks can help to clarify the problem of the feed-back: semantic → semiological. One of the more interesting aspects of the semio-narrative theory is to consider the narrative organization as a privileged manifestation of imaginary deep structures. It postulates that the 'meaning of life' includes *a constitutive lack of representation*, which is filled by narrative operations that find in this action their anthropological function. When it assumes that intentional narrative structures are syntagmatically controlled by delocalized categorical 'free' classemes, it is in fact postulating that the latter are 'deep' (e.g. life/death, man/god, nature/culture, etc.) and, though lexicalized by an embedding of the metalanguage in the object language, of an entirely different kind from the nuclear semes, semantemes and 'bound' classemes of lexical figures. We may say along with Thom that they are indeed categorizing and interoceptive with regard to their form, but 'prégnants' as regards their substance. That is why, they are so few in number and anthopologically universal.

According to this point of view, actantial structures syntagmatically unfold 'prégnance' trajectories (circulation of objects-values) which diffuse the initial 'free' 'prégnances' and transform them into delocalized 'bound' classemes defining semantic isotopies. The fact that these classemes can be identified at the lexical 'semiological' level is thus analogous to the investment of biological 'prégnances' in the perceptual saliences. In other words, there would be the same gap between the semantic 'free' classemes and the semiological 'bound' classemes as between the biological 'prégnances' and the perceptually salient forms. Narrative theory thus will have to show that :

(i) the actantial structures underlying the discursive-figurative manifestation, serve as 'channels' for the diffusion of imaginary 'prégnances';

(ii) there exists archetypal forms of circulation for these 'prégnances' as far as they are invested in objects-values;

(iii) the unfolding of the immanent deep semes can be done only by reconstructing the projection of the paradigmatic axis onto the syntagmatic one.

Thus, the semic 'prégnances' that are constitutive of the 'unconscious codes' would not be directly subjectivizable as such. Linked to the imaginary body (proprioceptivity and affectivity or 'thymicity'), they would be subjectivizable only *mediately* via the actantial events linking the subjects to their objects. The semantic/semiologic vicious circle pointed out by Brandt comes from the metalinguistic lexicalization of the 'prégnances' by terms like 'life/death', 'man/god', 'nature/culture', etc., which, though possessing no intrinsic subjective signification are nevertheless lexically definable. Semiotics joins here Metapsychology. Semiotic 'prégnances', those which are existentially most determinant, are not accessible to representation. They are unconscious, and this is precisely because they are at the same time maximally existentially determinant and minimally cognitively representable, and that they can be presentified only through their narrative dramatization in myths, tales, tragedies, novels, choreographies, etc.[1]

3.3 *Fundamental syntax and the semiotic square*

3.3.1 *Semiotic 'prégnances' and the substance of content*

In Greimasian theory, the semes of the semantic level (categorizing and interoceptive 'free' classemes) are treated in a Hjelmslevian manner. At the beginning, semes are a simple 'magnitude'

1 This is a narrative equivalent of the fact that there is no metalanguage. Just as, according to Wittgenstein, the grammatical form of the world cannot be represented, but only shown, so, according to Greimas, the embodied substance of the imaginary stance cannot be represented, but only actantially staged.

> whose semiotic existence is assumed prior to the analysis which will identify in it a discrete unit, and about which we only postulate the comparability with other magnitudes of the same order.[1]

They emerge from a substance articulated by the fundamental grammar, which contains two components: a *taxonomic* component dealing with the constitutive relations of the semes, and a truly *syntactic* component which transforms the taxonomic relations into *operations*; this latter is the key to the generative pathway since it underlies the *conversion* of logical operations into 'anthropomorphic' doings.

The articulation of the semantic substance is the condition for the *production* and the *apprehension* of meaning. It is based on the structural principle of the primacy of difference: 'we perceive differences, and thanks to this perception, the world 'takes form' in front of us and for us'.[2] This leads us to the semiotic square.

3.3.2 *The semiotic square*

As system of relations, the semiotic square is a 'primary taxonomic stance'[3] producing the elementary articulations of meaning. As 'the primary nucleus of an elementary morphology'[4], it belongs to the 'immanent level [...] where the narrative is situated and organized prior to its manifestation'.[5] 'Developed from a binary semic category'[6], it can be 'transformed into a constitutional semiotic model'[7] which is only 'the elementary structure of the signification used, as form, for the articulation of the semantic substance of a micro-universe'.[8]

> The isotopy of the terms of the elementary structure guarantees and posits in some ways the micro-universe as a unit of meaning, and allows us to con-

---

1 Greimas-Courtès, 1979: 168.
2 Greimas, 1966: 19.
3 Greimas, 1970: 163.
4 Ibid.: 164.
5 Ibid.: 158.
6 Ibid.: 160.
7 Ibid.: 161.
8 Ibid.

> sider, in the context of our axiomatic procedure, the constitutional model as a canonical form, as a starting point for a fundamental semantics.[1]

As *ab quo* instance, canonical form, formal universal, principle of semantic articulation, and generating structure of narratives, the semiotic square determines, according to Greimas, the structuration of meaning. It follows that the question of its mathematical nature is crucial for the theory.

But Greimas didn't deepen this question. For him, the semiotic square was endowed with all the semiotic 'energeia', but at the same time 'reified' in a trivial graphical icon which was only a 'visual representation of the logical articulation of a semantic category'.[2] We encounter here with Greimas the same problem as with Tesnière (see Sec. II.3): How to avoid the reification of the semiotic 'energeia' in the representation of semiotic form?

As a taxonomic and syntactic form prior to any semantic investment, the semiotic square develops a binary semic category s_1/s_2. In such a category, the terms are *positional values* defined by relations of junction (conjunction/disjunction) and of reciprocal presupposition (see Sec. 3.2.2). The relation of opposition s_1/s_2 is called *antonymy* and that which connects s_1 or s_2 with their semantic axis S (the category that unites them) is called *hyponymy* (or *hyperonymy* when it is orientated from S towards s_1 or s_2). The unfolding of this elementary structure into a semiotic square is necessitated by the conversion of the logical operations performed on the terms into 'anthropomorphic' doings. Indeed, at the 'anthropomorphic' level, the affirmation of a term s_1 opposed to s_2 generally happens via the *negation* of s_1. Further, 'a typology of relations is necessary, thanks to which one can distinguish the intrinsic features that constitute the category, from those foreign to it'.[3]

That is why, building upon Jakobson's distinction between *privative* oppositions of the type presence/absence ($A/{\sim}A$)[4] and *qualitative* oppositions of the type A/B, Greimas integrated contradictory terms ${\sim}s_1$ and ${\sim}s_2$ to the semic categories s_1/s_2. Whence the semiotic square:

---

1 Ibid.
2 Greimas-Courtès, 1979: 20.
3 Ibid.: 30.
4 If X is a term, ${\sim}X$ symbolizes non X.

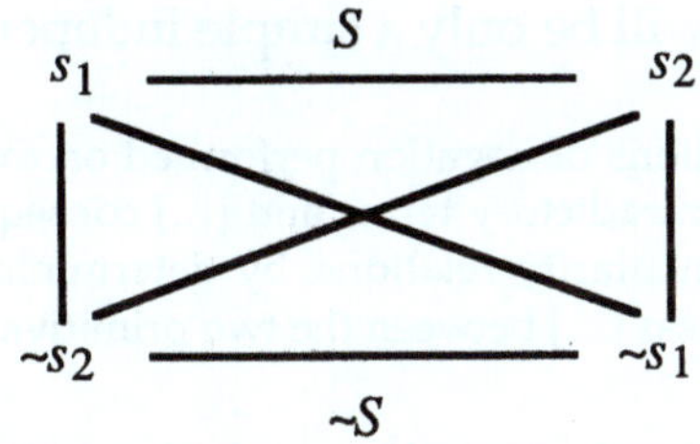

Three types of relations are constitutive of the square:

(i) hierarchical relations of *hyponymy* $s_1 \rightarrow S$, $s_2 \rightarrow S$, $\sim s_2 \rightarrow \sim S$ and $\sim s_1 \rightarrow \sim S$ (and the inverse relations of *hyperonymy*); they are *selection* relations in Hjelmslev's sense;

(ii) the categorical relations of *contradiction* $s_1/\sim s_1$, $s_2/\sim s_2$ and of *contrariness* s_1/s_2, $\sim s_1/\sim s_2$;

(iii) the relations of *implication* (unilateral presupposition, also called *hypotactic* relations) $\sim s_2 \rightarrow s_1$ and $\sim s_1 \rightarrow s_2$, which are also selection relations.

These relations can be regrouped into three *dimensions*:

(i) two *axes*: that of contraries $s_1 — s_2$ and that of subcontraries $\sim s_2 — \sim s_1$;

(ii) two *schemas* (dimension of contradiction): $s_1 — \sim s_1$ called (arbitrarily) positive, and $s_2 — \sim s_2$ called negative;

(iii) two *deixes* (dimension of implication): $s_1 — \sim s_2$ called positive and $s_2 — \sim s_1$ called negative.

We can see why the semiotic square is necessary for the definition of a semic category. The notion of a differential gap s_1/s_2 refers to an undefinable primitive relation whose logical nature 'remains undetermined'.[1] To ensure that this relation is really a relation of difference (conjunction/disjunction and reciprocal presupposition), it has to be *generated* as such. For that, we start from *A* and *B*, we negate them ($\sim A$ and $\sim B$), and we consider the *assertions* transforming $\sim A$ into *B* and $\sim B$ into *A* (implication). If there is a double assertion then

1 Greimas-Courtès, 1979: 30.

we can say that the A/B difference actually constitutes a semic category. If not, there will be only a simple independence:

> Two parallel operations of negation performed on the primitive terms [allow to] generate two contradictory terms and [...] consequently, two implications [establish] complementarity relations, by determining at the same time, the relation of contrariness [...] between the two primitive terms.[1]

Being only an elementary form, the semiotic square can of course be complexified. First of all, its constitutive relations can become terms of superordinate relations. For instance, there exists a relation of contradiction between the axes, and a relation of contrariness between the deixes. Further, the axis of contraries can be hypostatized on a term of the same level as s_1 and s_2 that V. Brøndal had called a *complex term* (s_1 and s_2), and the axis of subcontraries can be hypostatized on a term of the same level as $\sim s_1$ and $\sim s_2$ referred to by Brøndal as a *neutral term* (neither s_1 nor s_2). These terms play a fundamental role in mythic narratives.

The investment of the deixes by the *thymic* category (proprioceptivity) transforms them into *axiologies*: the positive deixis becomes euphorically connoted, and the negative deixis dysphorically. Axiology is the mode of existence of contents as *values* (both in the structural and anthropological sense). The axiological values are virtual and are actualized by the projection of the paradigmatic axis on the syntagmatic one. They become then *ideological*. According to Greimas, ideology is

> a permanent quest for value [and] an actantial structure which actualizes the values that it selects within the (virtual) axiological systems.[2]

3.3.3 *Logical inconsistency of the semiotic square*

As the semiotic square is the principle of articulation that mediates between the fundamental semantics and the anthropomorphic syntax (that is to say, between the Lévi-Straussian and Proppian sources of the semio-narrative theory), we must give the greatest importance to

1 Ibid.: 3.
2 Ibid.: 179.

the question of its formalness.[1] It is generally thought that the latter is of a logical essence. We would like to show why this cannot be the case.

Let us suppose that the square is a logical structure. Since it is constituted of relations of contradiction, contrariness and implication, it must be a Boolean algebra. Let us therefore consider a Boolean algebra of semes with 0 representing the empty seme and 1 the wholeness seme. It is natural to interpret a semantic axis as a relation of *orthogonality* in the Boolean sense. If s_1 and s_2 are the semes of a semic category, we have $s_1.s_2 = 0$ (orthogonality) and $s_1+s_2 = S$, where S is the seme of the semantic axis. By definition, we have $\sim s_1 = 1+s_1$, $\sim s_2 = 1+s_2$ (Boolean complementarity) and therefore,

$$\sim s_1+\sim s_2 = 1+s_1+1+s_2 = s_1+s_2 = S \text{ (since } 2 = 0)$$
$$\sim s_1.\sim s_2 = (1+s_1).(1+s_2) = 1+s_1+s_2+s_1.s_2 = 1+S = \sim S.$$

The alternative is therefore as follows. Either the $\sim s_1/\sim s_2$ axis *is not* a semic category, and the square is a heterogeneous structure. Or it is actually a semic category (contradictory of the s_1/s_2 category) and it should thus fulfill the condition of orthogonality $\sim s_1.\sim s_2 = 0$. But if $\sim s_1.\sim s_2 = 0$, then $\sim S = 0$ and $S = 1$, which implies that $\sim s_1 = s_2$ and $\sim s_2 = s_1$. In other words, a semiotic square can be Boolean *only if it is degenerate*, the hypotactic relations of implication collapsing into relations of equality.

Evidently, we can remove the condition of orthogonality and assume that the content of the semantic axis s_1/s_2 is the union $s_1 \vee s_2 = s_1+s_2+s_1.s_2 = S$, the product $s_1.s_2 \neq 0$ being the content of the complex term. The semantic axis $\sim s_1/\sim s_2$ would thus become a semic category defined by:

$$\sim s_1+\sim s_2 = s_1+s_2 = S+s_1.s_2,$$
$$\sim s_1.\sim s_2 = 1+S = \sim S.$$

From this perspective, the complex term of the axis of subcontraries would be the negation of the axis of contraries (the neutral term) and its content would be the symmetric difference between the content S of the contraries axis and the complex term $s_1.s_2$: to be non-degenerate a Boolean square must always have a neutral term and a complex term.

1 The semiotic square gave rise to many reflections, methodological as well as epistemological. See in particular SES 1976 and BGRS 1981.

But there are deeper reasons of principle, to reject the hypothesis of a logical formalness of the semiotic square. Logical algebras that we come across in the theory of formal languages presuppose very strong ontological hypotheses regarding the objects on which they can be applied.[1] They presuppose in particular that the objects are individuated and autonomous entities satisfying an identity principle, and that their relations are inherent. In model theoretic logic, the links between syntax and semantics result from this prior ontological hypothesis. Objects are defined in advance and syntax can be founded on a bracketing of denotation. Thus symbols are introduced for constants, variables, predicates, functions, operations, recursively constructed well-formed formulae, and logical operations (conjunction, disjunction, negation, implication, equality, quantifiers). These symbols are given an interpretation within specific mathematical structures, and thus is established a correspondence between closed formulas (without free variables) and statements. As for syntax, one resorts to the deduction of formulae from other formulae taken as axioms or as hypotheses, and as for semantics, one is interested in the truth value of statements. This strategy led to a very deep analysis of the relations between syntactic deductibility and semantic validity, and to extremely profound and highly non-intuitive results (e.g. Gödel's incompleteness theorems, Löwenheim-Skolem theorems, Tarski's truth theory, non-standard arithmetic and analysis, etc.).[2] But, however relevant these results can be, they cannot mask the symbolic lack of understanding of the paradigmatic order.

Semiotics is very different from logic.

(i) Structural units are not symbols but positional values. They are neither individuated nor autonomous, and exist only relationally.

(ii) Semantics is not a denotative interpretation of symbols but an articulation of a substance.

(iii) Syntax is not founded on a bracketing of denotation, but on an actantial conversion of the semantic articulation.

In a word, the relations between syntax and semantics are of a *radically different eidetic type* in the logical and the semiotic universes.

---

1 See sec. I.3.1, II.2.3 and II.4.1.

2 See, Petitot, 1979a.

Besides these reasons of principle, there are also technical reasons for rejecting a Boolean interpretation of the semiotic square. The first concerns the status of *negation*. In the logical interpretation, differences, which are really dynamical phenomena of differentiation, are reified into formal negations. This is a specially difficult problem, well understood by Greimas. As J. Courtès remarks, the axis of subcontraries $\sim S$ is not a true semantic axis, because $\sim S$ is interpreted as an 'absence of meaning'. Negation is a *metasemiotic* operation and it is impossible to make it the basis of the semiotic square without introducing inconsistencies. The semantic relations of contrariness and contradiction correspond to the two types of oppositions identified by R. Jakobson in phonology, the qualitative (polar) ones and the privative ones, respectively. And it is because the schemas $s_1 — \sim s_1$ and $s_2 — \sim s_2$ are *privative* oppositions that the axis of subcontraries $\sim S$ is not a true axis. A privative opposition 'presence/absence' is very different in nature from a contradiction.

As regards the privative oppositions, Alain de Libéra noted:

> We can [...] regard [the semiotic square] as a 'logical device' that produces privative oppositions from qualitative oppositions.[1]

Such a production can be understood only within the framework of a 'dialectical' conception of oppositions, as proposed by Arild Utaker quite accurately.[2] We will briefly comment on this paper.

3.3.4 *The aporia of privative opposition* [3]

The problem of understanding the relationship between qualitative and privative oppositions is particularly delicate. At the *functional* level, a privative opposition operates as a qualitative one. In Jakobson's view :

(i) every distinctive feature is defined in opposition to another feature, and

---

1 Libéra (de), 1976.

2 Utaker, 1974.

3 In this section, we refer to Utaker 1974, and the commentaries on this paper by Libéra, 1976, and Petitot, 1977b.

(ii) the presence of a feature excludes its opposite (principle of exclusion/participation, or of disjunction/conjunction).

If we consider a qualitative opposition A/B, we can say that the presence of A implies both the non-presence of B, which is in paradigmatic relation with A (principle of exclusion) and the *illocutionary negation* of B, B being excluded, non-selected. Thus, according to Jakobson, every qualitative opposition is accompanied by an *illocutionary* negation equivalent to a privative opposition.

Building on Jakobson's analysis, Utaker pointed out the ambiguity of the semiotic square. As the relation of contrariness is a qualitative opposition, it can also be interpreted as an illocutionary privative opposition (the selection of s_1 excludes s_2 with which it is paradigmatically related). As the relation of contradiction is a privative opposition, it can also be interpreted as an illocutionary negation. According to Utaker, the attribution of a logical formalness to the semiotic square is therefore to be rejected as inconsistent: the contradictions remain privative and have to be interpreted 'dialectically'. The question is therefore to treat in an eidetically homogeneous manner both qualitative and privative oppositions, considering them as two modes of a single phenomena of differentiation generating positional values.

As far as phonological oppositions are concerned, the morphodynamical models outlined in Sec. II.1 allow us to solve the problem. Let us return to the general model of a field $\sigma : W \rightarrow \mathcal{X}$ associated with the control of internal dynamics X_w by an external space W, and let us treat W as a substratum. The attractors of the internal dynamics X_w define competing local regimes, and the catastrophes induced by their destabilization categorize W. But we have seen that there are essentially *two* main types of catastrophes, namely catastrophes of *conflict* (competition of local regimes) and catastrophes of *bifurcation* (disappearance of a local regime). The former induce in W *symmetric* boundaries and the latter *asymmetric* ones. But whether it be symmetric or asymmetric, a boundary pertains to a *single* phenomenon of differentiation and thus, in so far as they are *external* features (in the sense of Sec. II.1.2.6), *all* distinctive features are of *the same eidetic type*. But, as *internal* features, they are either qualitative oppositions (conflicts) or privative oppositions (bifurcations).

Of course, as we already stressed, in phonology the principle of equivalence between the relational form and the organized substance (the emic/etic dialectic) is applicable. It is no longer the case in semiotics due to the lack of an etic level. The question is therefore of knowing whether the analogy with phonology proclaimed forcefully by the structuralist semioticians can be extended to the use of morphodynamical models as schemas of articulation for the semantic substance. We will address this problem in detail in another work.[#]

3.3.5 *The syntax of operations*

Fundamental syntax constitutes the second component of fundamental grammar. It transforms paradigmatic taxonomic relations into syntagmatically ordered syntactic operations, and serves as input to the anthropomorphic syntax. It thus represents the basic mediation between the two stages of the narrative grammar. As Greimas explains:

> 1. The narrative grammar contains an *elementary morphology* provided by the taxonomic model, and a *fundamental syntax which operates on the taxonomic terms* previously inter-defined.
> 2. Narrative syntax consists of operations that apply on terms likely to be invested with content-values; hence, it transforms and manipulates them, by negating and affirming them or by *disjoining* and *conjoining* them.
> 3. The syntactic operations, situated in the established taxonomic framework, are *orientated* and hence, can be predicted and computed.
> 4. Moreover, these operations are *arranged in series,* and constitute processes segmentable in *operational syntactic units.*[1]

Thus, the operations of the fundamental syntax can be considered as a *syntagmatic dynamization* of the terms defined paradigmatically by the semiotic square. While the taxonomic level of the semantic articulation corresponds to the primitive conditions of meaning apprehension, the syntactic level of operations correspond, in its turn, to the *ab quo* instance of discursive production.

> The examination of the conditions of meaning apprehension shows that if signification – to the extent that we seek to find it in the object – appears as an articulation of fundamentally stable relations, it is at the same time likely to be a dynamical representation as soon as we consider it as an apprehension or as a production of meaning by the subject. By taking into account this dynamical

\# Petitot, 1992.

1 Greimas, 1970: 165–166.

> aspect, we can establish a network of equivalences between the fundamental *relations* constitutive of the taxonomic model, and the projections of these relations, or *operations*, involving the terms already established by the very same elementary morphology, operations whose regulation would constitute the syntax. Thus, as a relation, contradiction serves at the level of the taxonomy, in the establishment of binary schemas; as an operation, it will consists, at the syntactic level, in a negation of one of the terms of the schema and an affirmation at the same time of its contradictory term. When such an operation takes place on terms already invested with values, it results in the transformation of the contents by negating those which are posited, and by making new asserted contents to emerge in their place. [...] Finally, the knowledge of the relational properties of the elementary structure – which are also that of the syntactic operations – prescribes the following: the operation of contradiction which by negating, for instance, the term s_1, posits at the same time the term $\sim s_1$, has to be followed by a new operation of presupposition that makes the new term s_2 to emerge and to combine with the term $\sim s_1$. Thus, the syntactic operations are not only orientated, but also arranged in logical series.[1]

The syntactic operations of assertion and negation are therefore transformations shifting from one term to another of the semiotic square. By conversion, they are reformulated as narrative statements of conjunction and disjunction between subjects of state [2] and objects of value.[3] The existence of an algorithm describing the path from an initial state to a final state shows a privileged concatenation of syntactic operations. A painstaking analysis of tales and myths prompted Greimas to opt for the '8-shaped' path:

$$s_1 \rightarrow \sim s_1 \rightarrow s_2 \rightarrow \sim s_2 \rightarrow s_1$$

1 Ibid.: 165.

2 At the actantial level, there are two kinds of subjects: subjects characterized by their state and subjects characterized by their doing (see the next section 3.4).

3 The key to the conversion is therefore the equivalence $s \equiv S \cap O$ between, on the semantic side, the affirmation of a seme s, and, on the syntactic side, a subject-object conjunction.

where, the acts of negation $s_1 \rightarrow \sim s_1$ and $s_2 \rightarrow \sim s_2$ actualize the relations of contradiction (schemas) and the acts of affirmation $\sim s_1 \rightarrow s_2$ and $\sim s_2 \rightarrow s_1$ actualize the relations of implication (deixes). We will note that in this syntagmatic dynamization of the semiotic square, the axes (relations of contrariness) are not actualized by any operation, while at the anthropomorphic level they correspond to the major event: the conflict between a subject and his anti-subject, that is to say, the *performance* manifesting the *polemical* dimension of the tale (see Sec. 3.4.4).

The movement from the taxonomic morphology to the syntax of operations raises a particularly difficult theoretical problem as it implies a transformation of the status of the entities considered. As it is a condition of possibility of the anthropomorphic conversion, we will call it a *preconversion*. It transforms the domains of a differentiated substratum (positional values defined by differences) into individuated and autonomous discrete units on which operations can act. In this sense, preconversion is the semantic equivalent of the *condition of discretization* explained in Sec. II.1.2.6.b.[1]

3.4 *Anthropomorphic syntax and the actantial theory*

Once we accept the possibility of a preconversion that projects the paradigmatic axis (taxonomic morphology) onto the syntagmatic one (syntax of operations), and transforms deep semes into discrete units, it becomes possible to *change the level*, and to move from fundamental syntax to anthropomorphic narrative syntax.

3.4.1 *The conversion*

Conversion is the change of level. What appears at the actantial level are essentially three types of relations, respectively between subjects and objects, subjects and anti-subjects, and subjects and senders.[#] The

---

1 This is what makes the question of the neutral-complex terms so delicate, for they correspond to positional values which violate the condition of discretization.

The French word 'Destinateur' is translated here as 'Sender'. It is an actant which motivates the subjects' intentionality and warrants the axiological va-

first group concerns episodes involving conjunctions and disjunctions between subjects and objects invested with values (objects of quest). They are described by narrative predicates of *state* # $S \cap O$ (subject-object conjunction) or $S \cup O$ (subject-object disjunction). The second group corresponds to the *polemical and conflictual* dimension of the tale, syntagmatizing the paradigmatic relation of contrariness (theory of performance). As regards the third group, it corresponds to the *contractual* dimension of the tale (the hero is initially motivated through a contract with a sender-mandator and, finally, sees his performance evaluated by a sender-awarder, in general a rewarder).

Conversion moves from fundamental grammar to anthropomorphic actantial syntax. The deep semantic categories which were only linguistic values are converted into axiological values invested in objects, the taxonomic relations into predicates of state, and the syntactic operations of the fundamental syntax into anthropomorphic *doings* governing these predicates. Thus we move from a formal structural syntax to an *event-based* syntax of action, of 'the meaning of life' and of 'desire'. We will see that the nature of this change of level is particularly problematic.

3.4.2 *Actants and the actantial model*

The narrative actants are involved in predicates, either static predicates of state (what are called *qualifications*), or dynamic predicates of process (what are called *functions* in the neo-Proppian sense). Whence a double analysis, qualitative and functional, subordinated to an actantial 'model'.[1] But contrary to the situation in logic where the objects are defined preliminarily and where predicates can therefore have truth values, in narratives the actants are pure 'bearers'. They are places progressively invested with values represented by the objects of quest with which they are connected. In other words, functions and qualifications are not 'properties' of predefined actants, but rather, events constituting progressively the 'existential content'

---

lues. We will also use 'Intender'. The neologism 'Destinator' could also be acceptable.

\# The French word 'Énoncé' is translated as 'Statement'. For convenience we will also use the expression 'predicate of state' instead of 'statement of state'!

1 See, Courtès, 1976: 61.

of 'formal syntactic units' prior to any semantic investment.[1] Narrative actants can be thought of in two ways. Firstly, as *syntactic* actants inscribed in narrative statements and programs. Secondly, as *functional* actants taking up the *actantial roles* of a narrative path. Indeed, in the course of the narrative, an actant is not only defined by its syntagmatic position. It is also defined by its *modal* investments. The conjunction of a syntagmatic position and a modal investment defines an *actantial role*. At the discursive level, the actantial roles are converted into *thematic roles* that make up the *actors* of the narrative.[2] Actors are discursive dramatis personae very different from narrative actants.

Narrative actants were conceived of by Greimas in a *case-like* manner, i.e., similar to the case notions of Tesnière and Fillmore. The statements of doing thus involved interactions of actantial places distributed by an 'organizing centre' (the verbal node). But, in comparison to the classical case-grammars, Greimas' actantial grammar is more fundamental. It reduces the verbal semantism to the single dimension of a syntactic *doing* prior to any semantic investment. A list of deep actantial events constitutive of the elementary functional statements shows that they fall into 3 classes:

(i) dynamical conjunctions $S \cap O$ and disjunctions $S \cup O$ between a Subject and an Object, that is to say, events of 'capture' and 'separation';

(ii) polemical conflicts between a Subject and an Anti-Subject;

(iii) transfers of an Object from a Sender to a Receiver.

More exactly, as the relations of conjunction and disjunction (predicates of state) are statical for Greimas, an event of conjunction $S \cap O$ expressed by the statement of doing $F(S_1 \rightarrow S_2 \cap O)$ (S_1 makes S_2 to conjoin with O) is called *realization*. On the figurative actorial plane, it yields an *acquisition* which can be either an *attribution* when it is transitive ($S_1 \neq S_2$) or an *appropriation* ('capture') when it is reflexive ($S_1 = S_2$). Similarly, an event of disjunction $S \cup O$ expressed by the statement of doing $F(S_1 \rightarrow S_2 \cup O)$ is called an *actualization*. On the figurative actorial plane, it yields a *deprivation* which can be either a *dispossession* when it is transitive or a *renunciation* when it is reflexive.

1 Greimas-Courtès, 1979: 3.
2 Ibid., item 'Actant'.

Greimas' actantial model is got by placing the basic actants (Subject and Object, Sender-Intender and Receiver, Helper and Opponent) within a single schema as given below:

$$\begin{array}{ccccc} \text{Sender} & \longrightarrow & \text{Object} & \longrightarrow & \text{Receiver} \\ & & \uparrow & & \\ \text{Helper} & \longrightarrow & \text{Subject} & \longleftarrow & \text{Opponent} \end{array}$$

In our opinion, this model presents a gap as regards the basic relation of *conflict* between a Subject S_1 and an Anti-Subject S_2 which is the core of the Subject's performance. Greimas resolves this difficulty by observing that the polemical structure of the narrative is due to the intertwining of two antagonistic narrative pathways (that of the hero, and that of the villain). But these pathways are not independent. They are linked by the performance and that is why the latter should become part of the actantial model. The S_1/S_2 conflict is the anthropomorphic conversion of the taxonomic relation of contrariness. It makes the conjunction $S_1 \cap O$ the counterpart of the disjunction $S_2 \cup O$, that is to say, the realization of a *polemical transfer* $S_2 \rightarrow O \rightarrow S_1$ according to the schema:

$$S_2 \xrightarrow[\substack{S_2 \cup O \\ \text{(Dispossession)}}]{\qquad\qquad} O \xrightarrow[\substack{S_1 \cap O \\ \text{(Appropriation)}}]{\qquad\qquad} S_1$$

Thus we would have two different types of transfers:

(i) contractual transfers:

$$\begin{array}{ccc} D & \longrightarrow O \longrightarrow & d \\ \text{Sender} & & \text{Receiver} \\ \text{(Renunciation)} & & \text{(Attribution)} \end{array}$$

where the object transmitted is, for instance, a modal or a cognitive value and where the renunciation of the intender is not a real deprivation (what is called a participative communication), and

(ii) polemical transfers (of the type dispossession/appropriation) where the object transmitted is an object of desire, for instance, a pragmatic value.

Considered at a mere abstract level, the elements of the actantial syntax may appear rather trivial. But it is no longer the case if we tackle the problem of their *formalization*. If we suppose *a priori* – as is done by most of the linguists and semioticians – that in semio-linguistics formalization is only a matter of symbolic representation of units, relations and primitive structures, then we would be led to think of the narrative syntax as a mere combinatorics of narrative statements, and hence to criticize its triviality and redundance. Greimasian theory will then be blamed as a grindmill that crushes the richness of discursive inventiveness. But the situation changes completely as soon as we seriously take into account the fact that Greimasian theory concerns a syntactic conversion of semiotic 'prégnances' and refers ultimately to anthropological substantial universals. The Subject/Object relation should be conceived as an *intentional* and *teleological* one. As Greimas explains, a relation of intentionality $S \rightarrow O$, dynamic, orientated, asymmetric and irreversible 'has to be necessarily postulated, prior to the constitution of any actantial syntax'.[1]

Such an intentionality is a condition for the *semiotic existence* of actants: semiotically, subjects and objects exists only as a function of their junction; prior to their junction they are only virtual places; they are actualized by a relation of disjunction making the subject in quest of a desired object. As 'quest', a narrative must henceforth be conceived of as a program of realization (conjunction $S \cap O$):

> by realizing its narrative program, the subject realizes the value which was only aimed at, as well as he 'realizes' himself.[2]

In this regard, formalization of the actantial syntax has to model the intentional dynamics underlying the relations of disjunction/conjunction, and of virtualization/actualization/realization.

---

1 Greimas-Courtès, 1979: 402.

2 Ibid.: 139. Greimas plays on the two meanings of 'realize': to make real something virtual and to fulfill a destiny.

Besides the polemical dimension Subject/Object/Anti-Subject, narratives also include a contractual dimension of 'communication' wherein Intenders play a prominent role. First, before being 'realized' by way of some performancial tests, the hero has to become a *competent* subject, and for this he must acquire (by way of gifts or through other trials) *modal* competences (powers, wills, attainments, etc.). Secondly, his performances (which reequilibrates an initial lack) are the consequences of a *contract* with an Intender (for instance a king) representing a social order. Thirdly, these performances have to be evaluated and approved both pragmatically (award) and cognitively (recognition) by an Intender.

> Often posed as belonging to the transcendent universe, the Sender-Intender is the one who communicates to the Receiver-Subject (pertaining to the immanent universe) not only the elements of modal competences, but also the whole of the sought-for values; he is also the one who rewards the result of the performance of the Receiver-Subject. From this point of view, we could therefore draw an opposition, in the narrative schema, between the (initial) Sender-mandator and the (final) Sender-awarder.[1]

While communicating his modal competence as well as his 'sense' of values to the Subject, the transcendent Intender does not become deprived of them. His communication is participative. It is not the case in the immanent universe of the subjects where everything happens as if the 'inventory' of values were limited, closed, and 'conservative', each junction $S_1 \cap O$ implying a corresponding disjunction $S_2 \cup O$ (polemical transfer).

3.4.3 *The objects of value and their circulation*

In Greimasian theory, 'the narrative organization of values' is the basis of narrative.[2] Subjects exist semiotically only by their junctions with objects, and objects only as support of values 'as a space of fixation [...] of values determinations'.[3] This equivalence between values in the structural and axiological sense is a consequence of the basic hypothesis according to which only the syntactic 'representa-

1 Greimas-Courtès, 1979: 95.
2 Greimas, 1973a: 35.
3 Ibid.: 15.

tion of the imaginary stance' enables 'to imagine the apprehension of meaning and the manipulation of significations'.[1]

> The object is a syntactic concept, an end-term of our relation to the world, but at the same time one of the terms of the elementary statement which is a semiotic simulacrum representing in the form of a scene, this relation to the world.[2]

The apprehension of meaning never encounters the object as such but only the values it bears : the lexeme 'appears deceptively ('en trompe l'oeil') in the place marked for the object'.[3] We again come across the idea that discursivization is a clothing that veils deep significations.

Inscribed at the semantic level and syntactically converted, the values drive the narrative pathways. They are of several types. They can be either modal or descriptive, and the latter can further be objective (realized according to the 'to have' mode) or subjective (realized according to the 'to be' mode). The values aimed at by the subject's intentionality are called *basic* values and are opposed to *use*-values (instrumental values). The descriptive values are of a pragmatic nature. To be recognized as values by the subject, their *pragmatic* dimension has to become an internal referent of a *cognitive* dimension. The cognitive doing of a subject is articulated by truth modalities (true/false/secret/lie) and functions generally as the stake in a *fiduciary* contract between the Sender-Intender and the Receiver. By means of his persuasive doing, the Intender manipulates the Subject who in turn accepts the contract and its values, at the risk of being deceived.

As we mentioned above, in the immanent universe, the values driving the subjects' intentionality constitute a 'closed axiological universe'.[4] There are two types of circulation of values. On the one hand, they are exchanged in the 'conservative' universe where the polemical transfers prevail, and on the other hand, they communicate with the transcendent universe in participative communication. The fact that they circulate in a closed system explains their polemical dimension. The fact that they should be certified by the intender

1 Ibid.: 16.
2 Ibid.
3 Ibid.
4 Ibid.: 23.

explains their contractual dimension. The contractual sender represents the *paradigmatic* system of the fundamental semantics.[1] The projection (preconversion and conversion) of the paradigmatic axis onto the syntagmatic one is therefore manifested as the relation between the contractual and performancial sequences. Taken as a narrative program for the subject, the contract transforms the latter into a *performer*-subject (depending on the acquisition of modal competence), ensuring the mediation between the (paradigmatic) system and the (syntagmatic) process.[2]

3.4.4 *Narrative statements and programmes, tests, and polemical structure*

Through anthropomorphic conversion the fundamental grammar generates narrative predicates, either of state, or of doing. These statements are linked up in elementary syntagms referred to as *narrative programs*. Besides the contractual sequences, the most typical and most important narrative programs are the *tests*. A test is a polemical transfer (appropriation/dispossession) which is a reflexive conjunction:

$$F(S_1 \rightarrow S_1 \cap O) \equiv F(S_1 \rightarrow S_2 \cup O)$$

Canonically, there are three successive tests in a narrative (if one does not take into account their repetition for the sake of emphasis): the qualifying test, the decisive test (performance), and the glorifying test. The first two are pragmatic and the third cognitive. In relation to the basic narrative programme (performance), the qualifying test is a use-program leading to the acquisition of modal competence. The glorifying test concerns the recognition of the Subject's performance by the Awarder-Sender. In general the Subject convinces the Sender about his performance by displaying a truth marker or a verificatory mark (a trophy, an evidence showing up the traitor, etc.).

The 'most typical unit of narrative syntax'[3] is the performance whose polemical character converts the relations of contradiction and contrariness.

---

1 See, Courtès, 1976: 99.

2 The system/process opposition is a Hjelmslevian variant of the paradigmatic/syntagmatic one.

3 Greimas, 1970: 173.

> If we accept that the anthropomorphic representation of contradiction is of polemical nature, then the syntagmatic sequence – which corresponds to the transformation of the values of content resulting, at the level of the fundamental grammar, from the operations of negation and assertion – will appear as a series of narrative statements whose semantic restrictions will confer on it a feature of confrontation and struggle.[1]

According to Greimas, this central narrative program presupposes:

(i) the confrontation of two antagonistic subjects S_1/S_2 (or $S/{\sim}S$) associated with two opposite doings;

(ii) the equivalence between the operation of negation and the polemical domination of one subject over the other;

(iii) the equivalence between the orientation of these operations and the choice of the dominant subject;

(iv) the surface representation, by means of two complementary narrative statements, of the 'dialectical' equivalence between the negation of a term and the affirmation of the contradictory term followed by the assertion of a contrary term.

Governed by an order of implication $NS_3 \rightarrow NS_2 \rightarrow NS_1$ (NS = Narrative Statement) corresponding to the orientation of the deep syntactic operations, a performance would thus link up the following three narrative statements:

NS_1 : S_1/S_2 confrontation (syncretism of two modal statements proper to each of the subjects);

NS_2 : $S_1 \rightarrow S_2$ domination, with the negation consisting

> in the transformation of a virtual state into an actualized one, or, what amounts to the same, the substitution of the *MS* (modal statement) of will by the *MS* of existence, of the desire for domination by domination;[2]

NS_3 : $S_1 \leftarrow O$ attribution, where the assertion is equivalent to the attribution of the value-object to the subject.

We will focus a moment on this description of performance. Let us consider a standard narrative (such as the myth of St. George) where

1 Ibid.: 172.
2 Ibid.: 173.

a hero saves a princess attacked by an anti-subject (traitor, villain, trickster, dragon, etc.).

> The circulation of values, interpreted as a series of transfers of object-values can adopt two pathways :
> $$F(d_1 \rightarrow O \rightarrow {\sim}d_1) \rightarrow F({\sim}d_1 \rightarrow O \rightarrow d_2)$$
> which in the special case of Propp's Russian tales can be interpreted as follows: the society (d_1) experiences a lack, the villain (${\sim}d_1$) attacks the daughter of the king (O) and carries her away in order to hide her (d_2).
> $$F(d_2 \rightarrow O \rightarrow {\sim}d_2) \rightarrow F({\sim}d_2 \rightarrow O \rightarrow d_1) \ (2)$$
> which would mean: the hero (${\sim}d_2$) finds in some place (d_2) the king's daughter (O) and returns her to her parents (d_1).[1]

In such a myth or a fairy tale, the basic opposition is between Culture and Nature. Culture is represented by a collective actant (society) personalized by an Intender (king) localized in a town (in the pictorial representations of the myth of St. George, the town acts as a figurative metonymy for Culture).[2] Nature is often neither represented nor personalized, but only localized in a no-man's land exterior of the town (cave, etc.). Each of these universals belonging to the transcendent universe (the king being a figure of the Intender) 'emanates' in the immanent universe a substitute, the hero (S) for culture, the villain or the monster (${\sim}S$) for Nature. This 'emanation' develops along the hyperotactic relations of the deixes of the semiotic square:

Culture (C): Town (T)	———	Nature (N): Elsewhere (E)
↓		↓
S (Hero : Subject)	———	${\sim}S$ (Villain-Dragon: Anti-Subject)

Let us comment on these different levels of Greimas' description.

1 Ibid.: 177. The letter *d* refers to the deictic interpretation of the semiotic square, which we will discuss in the next section.

2 See, Petitot, 1979d.

1. The narrative is initially *axiologically polarized* by the thymic investment of the semiotic square (euphoric deixis/dysphoric deixis). Polarization introduces an asymmetry between values and orients the syntagmatic unfolding of the fundamental taxonomy.

2. The narrative begins with the establishment of the initial schema $T/{\sim}S$. The contradiction negates T by affirming its contradictory ${\sim}S$ and 'must be followed by a new operation of presupposition that yields a new term (E) that joins up with the term ${\sim}S$'.[1] But we shall note that the anthropomorphic conversion of the 'logical' path $T \rightarrow {\sim}S \rightarrow E$ complicates this operational conception of contradiction. On the one hand, by anthropomorphizing the term 'absence' of the privative opposition $T/{\sim}S$, by making the Anti-Subject ${\sim}S$ the 'illocutionary negation' of Culture (see, Sec. 3.3.4), and by personalizing in the king (D[2]) the seme 'Culture' and the collective actant 'Society', it transforms contradiction into a *polemical transfer* $D \rightarrow O \rightarrow {\sim}S$, according to the equivalence:

$$\begin{array}{ccccc} T = C & & \xrightarrow{\hspace{6em}} & & {\sim}T = {\sim}C \\ D & \xrightarrow[D \cap O]{} & O & \xrightarrow[\substack{{\sim}S \cap O \equiv \\ D \cup O}]{} & {\sim}S \end{array}$$

We see from this example how conversion involves the substitution of semes by predicates of state, and in particular, of the 'initial' seme s_1 by an actantial conjunction $D \cap O$. If we suppose a permanence for the actants (an abstract principle of identity), and the allocation of the semantic charge to the object, then we can interpret the semic negation constitutive of the contradiction $s_1/{\sim}s_1$ by the syntactic transformation $D \cap O \rightarrow D \cup O$. The negation $D \cap O \rightarrow D \cup O$ *actualizes* O as value (initial lack). O becomes thus a value of quest for the hero who has contractually stood up for the king (the social order), and triggers its program of realization. If we further hypothesize that the immanent universe of axiological values is a *closed* universe, then the disjunction $D \cup O$ that institutes the lack is equivalent to the conjunction ${\sim}S \cap O$ and it would be right to treat the

1 Greimas, 1970: 165. See, Sec. 3.3.5.

2 We will keep the initial D of 'Destinateur' for symbolizing the Sender-Intender.

anthropomorphic conversion of the contradiction as a polemical transfer $D \rightarrow O \rightarrow \sim S$.

3. Once we accept these hypotheses, we can interpret the constitutive implication of the negative deixis ($\sim d_1 \rightarrow d_2$) as a non-polemical transfer, as a *gift*: $\sim S \rightarrow O \rightarrow E$ (renunciation/attribution). This gift anthropomorphically converts the assertion following the negation.

4. But when one tries to develop these hypotheses up to performance, one stumbles upon a number of difficulties.

(a) Firstly, the king personalizing the social order (Culture) functions simultaneously as an Intender and as a dispossessed Subject, with the program of realization of the hero (liquidation of lack) being for him a restitution/reparation.

(b) By asymmetrizing the deixes of the semiotic square, the axiologization of values produces not only a syntagmatic orientation but also an asymmetry between the narrative pathways of the Subject and the Anti-Subject. In a symmetrical narrative, the Elsewhere would be represented by the town of an another kingdom, and the contrariness d_1/d_2 would confront, for instance, two kings (one D 'good and just' and the other $\sim D$ 'wicked and unjust'). The narrative path of the Subject would be then symmetrical with that of the Anti-Subject: S would tear off O from $\sim D$ in order to restitute it to $D = T = C$ (polemical transfer):

$$\sim D \xrightarrow[\sim D \cap O]{} O \xrightarrow[\substack{\sim D \cup O \equiv \\ S \cap O}]{} S$$

followed by the gift $S \rightarrow O \rightarrow D$). In this case there wouldn't be any conflict between the Subject and the Anti-Subject, the performance being situated on the axis of *contradictories* $S/\sim D$, and not on the subcontraries' axis $S/\sim S$. Generally,[1] the relation of contradiction $\sim D \rightarrow S$ is 'factorized' through $\sim S$ by means of a delegation $\sim D \rightarrow \sim S$:

1 See, for instance, a heroic narrative like *Ivanhoe*.

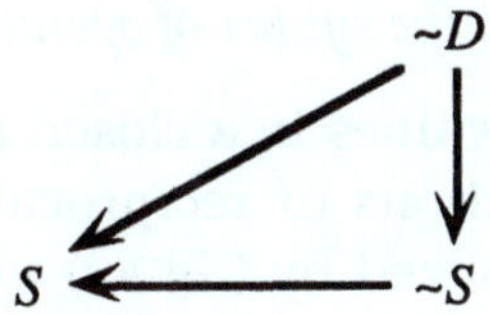

Hence a double conclusion:

(i) The condition of possibility of the performance is a hyperotactic factorization :

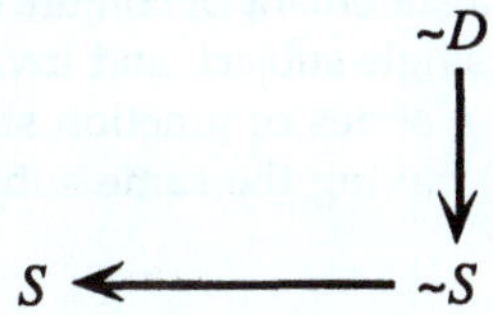

(ii) This factorization does not concern the object but the subjects, the 'victory' of S over $\sim S$ being equivalent to a transfer $\sim D \rightarrow O \rightarrow S$.

The main theoretical problem is that of such an equivalence.

(c) The ambiguous shift of the polemical category $S / \sim S$ (which converts the subcontraries axis) onto a schema of contradiction, results from the Greimasian principle according to which conversion is applied on the fundamental syntactic operations and can therefore affect only the *junction* relations S–O which actantialize the deep semes. But in order to account for performance as the decisive test, it is in fact necessary to modify Greimas' idea that the semantic content of a subject is defined only by his junctions with value-objects. At every step of the narrative, S can also be 'negated' *as such*. When an Anti-Subject $\sim S$ is negated by a performer Subject S, the semantic content of $\sim S$ diffuses as a pregnance from $\sim S$ to S. In other words, in a performance, while aiming at a value-object, the subject also aims at the *being* of the anti-subject possessing the object. The desire for the object is also the desire fo the anti-subject's desire for this object, what René Girard called a *mimetic desire*. It would be therefore necessary to schematize performance as an equivalence between a mimetic identification $S \approx \sim S$ and a polemical transfer $\sim S \rightarrow O \rightarrow S$.

3.4.5 Topological syntax and the syntax of operators

The circulation of object-values in a closed axiological universe leads, as we have seen, to relations of reciprocal presupposition between predicates of state symbolized by $(S_1 \cup O) \equiv (S_2 \cap O)$, or $S_1 \cup O \cap S_2$.[1] For these relations, Greimas proposed the term *paradigmatic junction.*

> We propose to use the term *paradigmatic junction* for the logically necessary concomitance of two statements of conjunction and disjunction affecting two distinct subjects. However, as far as narratives can be considered as series of narrative states, where a statement of conjunction presupposes a statement of disjunction involving a single subject, and inversely, we can reserve the name *syntagmatic junction* for a series of junction statements (conjunction and disjunction, or conversely) having the same subject and linked by a relation of simple presupposition.[2]

Whence the description:

	Paradigmatic
Syntagmatic	$S_1 \cup O \rightarrow S_1 \cap O$ ↓ ↓ $S_2 \cap O \rightarrow S_2 \cup O$

This formulation shows the reciprocal presupposition of two narrative programs in competition (those of the subject and the anti-subject), 'the solidarity between which is guaranteed by the concomitance of the functions.'[3]

In the actantial conception previously sketched, the syntactic operations were converted into predicates of doing governing predicates of states, and we could, for instance, describe performance as a coupling of a reflexive realization (appropriation) $F(S \rightarrow S \cap O)$ with a transitive virtualization (dispossession) $F(S \rightarrow {\sim}S \cup O)$. In this perspective, subjects are intentional subjects. But, elsewhere (perhaps because the notion of intentionality was not syntactically expressible), Greimas proposed to introduce an operational *meta-subject* Σ which is

1 See Greimas, 1973a.
2 Ibid.: 15.
3 Ibid.: 25.

'meta-tactic in relation to the subjects of predicates of state'[1] and which would be responsible for the syntactic doings affecting the predicates of state. Σ can uniformly substitute the intentionality of the subject-actants by an actantial *syncretism* $\Sigma \equiv S$ or $\Sigma \equiv \sim S$. For instance, performance will be expressed by the statements $F(\Sigma \equiv S \rightarrow S \cap O)$ and $F(\Sigma \equiv S \rightarrow \sim S \cup O)$. This is not just a mere notational variant. In fact, the introduction of Σ transforms all the transfers into *communicational* transfers, and performancial appropriations into attributions. Everything would appear as if at the level of the invisible meta-subject, the polemical confrontations between the subjects and the anti-subjects were only a dramatization of pre-programmed events. Obviously, Σ could be identified with the transcendent Intender D representing the paradigm of values. But this would make the actants some sort of manipulated 'puppets'. In fact, what Greimas had in mind was a transformation of actants into pure places crossed by the circulation of value-objects:

> the actants are conceived no longer as operators but as places where the value-objects can be located, places where they can be drived, or from where they can be withdrawn.[2]

We thus get a *topological syntax of value-objects*,

> a topological interpretation of the narrative according to which the displacements of objects alone would suffice to account for its organization, the subjects being only places of their transfers.[3]

We are therefore committed to model a double determination of actants as intentional subjects of doing and as deixes or places.

In the topological syntax actants lose their operational character. They are no longer 'subjects endowed with a specific virtuality of doing, able to accomplish the predicted operation of transfer'.[4] Their intentionality and operationality have therefore to be restablished.

---

1 Ibid.
2 Greimas, 1970: 176.
3 Greimas, 1973a: 35.
4 Greimas, 1970: 178.

This is done through the meta-semiotic level of *modalities* which 'should be developed for explaining the transfer of values.'[1]

Astonishing indeed is Greimas' circuit that starts from intentional subjects, moves to actants-deixes shifting their intentionality to the meta-subject Σ, and then returns to the subjects by way of modalization!

3.5 *The gap between fundamental syntax and anthropomorphic syntax*

As pointed out by several commentators, in spite of its rigor, Greimasian theory was not able to clarify the real eidetic status of the conversion transforming the semantic constitutional model to the modalized actantial one. In his *Introduction to Narrative and Discursive Semiotics*, J. Courtès emphasizes this point: 'It is unlikely that there exists a direct route from the elementary structure of signification to the surface syntactic distribution'.[2]

> Even the syntactic modalities of wanting-to and/or knowing and/or power – logically prior to the surface narrative statements, and therefore situated at a deeper level – do not allow us to connect the 'surface grammar' with the 'deep grammar'.[3]

This bias in the understanding of conversion does not matter for a conceptual-descriptive theory that makes use of undefinable primitives as well as principles of parallelism: it is sufficient to assume that anthropomorphic syntax is a *praxic representation* of logical syntax, introducing modalities associated with doing. But the bias matters a lot for a theory that seeks to be deductive and formal. As Courtès observed, the question then becomes that of the 'formal relation' between the fundamental and the surface levels.

The problem is that there exists *a formal gap between logical relations and syntactic events*. As the narrative actantial theory is case-based (even deeper than case-grammars), its real nature can be thought of as *localist* (see Sec. II.6): actants have to be modeled as positional proto-actants.

---

1 Ibid.: 178.
2 Courtès, 1976: 81.
3 Ibid.: 84.

The introduction of a localist conception allows us to explain the ambivalence between subject-actants and deixis-actants and also to project onto actantial syntax a part of the *spatio-temporal programming* that for Greimas remains confined to the discursive-figurative component: just as actants are pre-figurative and are not to be confused with actors, so there exists *pre-figurative places whose articulation is constitutive of actantial syntax* (as topological event-based syntax). If the latter cannot be deduced from a conversion of the logical syntax, it is simply *because the fundamental syntax is itself topological and not logical*.

In order to fill the gap that separates the fundamental syntax from the anthropomorphic syntax, we have to project the topological and event-based dimensions onto deep structures, and reformulate the semiotic square in such terms.

3.6 *Modalization and actantial roles*

Topological syntax treats the subject-actants as deixes and as *virtual* subjects of doing. In order to become operational, they have to be *modalized*. As acquisition of competences, modalization is prior to performance. It establishes a parallel between the continuous semiotic 'filling up' of the subjects with axiologized values (semiotic 'prégnances') and the modal evolution of their doing.

Modalities concern either doing (modalities of wanting-to or will, having-to or duty, being-able-to or power, and knowing-how-to or knowledge), or being (truth or 'veridictive' modalities).[#] Modalization of doing is distributed along the actantial model: wanting-to concerns the Subject/Object axis, having-to the Sender-Intender/Receiver axis, and knowing-how-to the Helper/Opponent axis. Wanting-to institutes the subject's desire. The will of the Intender is manifested by the having-to of the subject.[1]

Modalization of doing enables us to define *actantial roles*

\# We use here the standard translation for Greimas' modalities: *vouloir*, *devoir*, *pouvoir* and *savoir*.

1 But one can well say that the subject's will is also a self-referential duty (see, Greimas-Courtès, 1979: 96).

> by the *position* of the actant in the logical concatenation of narration (syntactic definition) and by its *modal investment* (morphological definition), thus rendering possible the grammatical ordering of narrative.[1]

Let us compare this definition with the entry 'Narrative pathway' given in the *Dictionary*:

> Once inscribed in the narrative path, the syntactic subjects can be defined – for each *NP* (narrative program) – by the position they occupy [...] in the pathway and by the nature of the value-objects with which they join up. [...] We will call 'actantial role' this double definition of the syntactic actant, i.e., by its position and its semiotic being.[2]

We can notice here the ambiguity of the definition of the subject's 'semiotic being'. At first, it is constituted of the acquisition of values determined by the fundamental morphology and realized by intentional subjects of quest after having been actualized by the fiduciary contract. But later, it is also constituted of the modal values.

Despite this difficulty, the actantial roles are interesting because they facilitate the movement from *syntactic* actants to *functional* actants which will be *actorialized* in the discursive-figurative component through the addition of *thematic roles*. This passage from the syntactic to the functional level is in some ways a passage *from the local to the global*. Syntactic actants are abstract 'local' actants converting the fundamental 'logical' operations, while functional actants take on several actantial roles, subsume global narrative pathways, and belong to *the narrative schema* (see, Sec. 3.9).[3]

3.7 *'Veridiction' and manipulation*

The modalities that we have discussed so far are modes of doing and are a matter of competence. But there are also modes of being, called truth or 'veridictive' modalities.

---

1 Greimas, 1973b: 165 (see also, Courtès, 1976: 76).
2 Greimas-Courtès, 1979: 242.
3 See, ibid.: 243.

As modalization of being, veridiction [#] is of a cognitive order. It articulates the knowing-that on being by unfolding the epistemic category being/seeming according to the semiotic square:

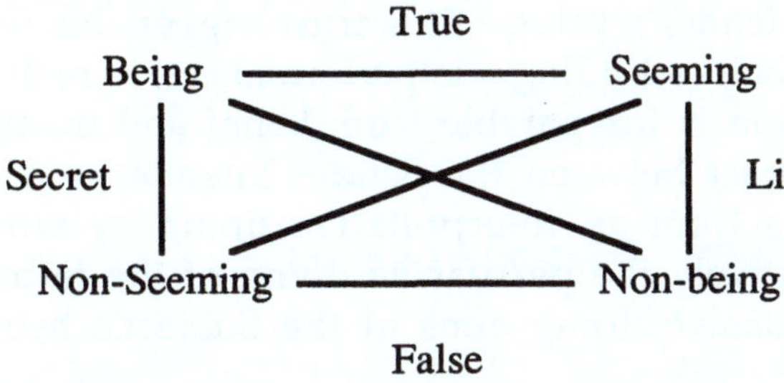

where the being/non-being schema is that of immanence and the seeming/non-seeming schema is that of manifestation.

> The category of veridiction is presented as the framework within which takes place the epistemic cognitive activity which, with the help of different modal programs seek to attain a veridictive position that can be approved by a final epistemic judgement.[1]

Veridiction plays a key role in the contractual component of narratives: the recognition of the hero is, for instance, a cognitive approval made by the Sender-Awarder warranting the truth of the values; the villain's exposure transforms the negative deixis of 'Lie' into the 'False' subcontraries axis; tales like *Cinderella* or *The Donkey's Hide* transform in the same way the positive deixis of 'Secret' into the 'True' contraries axis. Veridiction also appears in the performancial component. Let us consider for instance the *simulated* tests where the Anti-Subject is a representative of the Intender (Jacob's struggle with the Angel). They transform the 'True' axis into the 'Secret' deixis. On

---

\# 'Veridiction/veridictive' is a theoretical neologism constructed on the paradigm 'prediction/predictive' for referring to the narrative strategies concerning truth. 'Veracity/veracious' or 'veridicality/veridical' are not satisfactory. As for the veridictive modalities we could eventually speak of 'truth modalities' just as we speak of 'truth values' in logic. But we cannot avoid the term 'veridiction'.

1 Greimas-Courtès, 1979: 419.

the contrary, *deceptive* tests transform the 'False' axis into the 'Lie' deixis, and so on.

Moreover, veridiction appears in a still more constitutive manner in the initial contract where the subject of quest accepts the truth of the Intender's values. This trust triggers his wanting-to and having-to modalities leading, via performances, to realization. At this level, veridiction is inseparable from belief and manipulation. The fiduciary contract between the Sender-Intender and the Receiver-Subject results from an interpretative doing by which the latter evaluates positively the persuasive doing of the former. When the Sender's persuasive doing aims at the Subject's being, it has the function

> of conferring on the semiotic activity – which can be received only as a manifestation – the status of immanence, that is of inferring the noumenal from the phenomenal.[1]

The Subject's interpretative doing leads him to believe, doubt or reject the immanence. When the Sender's persuasive doing aims at the Subject's doing, it is a matter of *manipulation*. Manipulation is articulated on the semiotic square of iterated doing:

$$F(S_1 \rightarrow F(S_2 \rightarrow \ldots)).^2$$

In short, veridiction aims at accounting

> for this extraordinary 'game of masks' [...] which constitutes one of the essential axes or the narrative imaginary instance.[3]

3.8 *Discursivization and figurativization*

In order to complete our panorama of Greimas' theory, let us take a brief look at the discursive-figurative component.

---

1 Ibid.: 275.

2 Regarding manipulation, see BGRS, 1977.

3 Greimas, 1973b: 166. For other remarks on veridiction, see, Brandt, 1982a, b, and Petitot, 1982e.

3.8.1 *The function of nuclear semes: discursive configurations and themes*

In our presentation of the fundamental grammar and anthropomorphic syntax, we have emphasized the fact that

(i) the interoceptive and categorical semes of the taxonomic morphology were 'free' classemes resulting from the articulation of semiotic pregnances (see 3.2.3 and 3.3.1), and

(ii) they selected 'bound' classemes for sememes whose nuclear semes (semiological and non-semantic) were assigned to the discursive-figurative component. This latter is evidently the most manifest and profuse component of narratives. It displays their 'visible' richness.

The exteroceptive nuclear semes constitute *figures* paradigmatically organized in *discursive configurations* (a figure such as 'the sun' is related with satellite figures like 'rays', 'warmth', 'light', etc.). Syntagmatically, they unfold in *figurative pathways*. Discursive configurations

> appear as some sort of micro-narratives having an autonomous syntactico-semantic organization, capable of being part of larger discursive units, and acquiring functional significations corresponding to the global device.[1]

Their problematic is related to that of *motifs* and *stereotypes*. In other words, they are elements of the semio-cultural codes and lexicons that provide the narratives with their substance.

Now, this substance being constrained by the actantial organization, there must be, according to Greimas, *sememic* structures involved in the syntax of dramatis personae. They are called *themes*. In the form of thematic roles, themes are selected by the actantial organization and taken over by the *actors* of the narrative. A thematic role reduces a discursive configuration to a single figurative pathway and takes it through a competent agent. And as far as the thematic roles are selected by the actantial roles, there exists a tension between the narrative and the discursive, that is, between, on one side, the syntactic actants of the narrative programs and the functional actants of the narrative pathways, and on the other side, the proliferation of

1 Greimas-Courtès, 1979: 58.

discursive configurations and figurative themes scattering 'the intertextual migrations of the motifs'.[1]

Discursivity thus integrates the semiological level with the 'canonical grammatical forms' generated by the actantial grammar.

3.8.2 *Thematic roles and actors*

Thematic roles indicate how the actantial syntax appropriates discursive configurations. As Claude Chabrol observed:

> the notion of thematic roles organizing and selecting figurative pathways in discursive configurations unfolded from lexematic figures, must form a plan of mediation and articulation between the phrase and the text structures.[2]

They are assumed by the actors whose semantic investment they define. Now, to the extent the actors are governed by the actantial syntax, they constitute the key mediation between the narrative and the discursive levels.

> [They] have a double mission : on the one hand, they support the narrative structure [actantial roles] distributing the fundamental functions along the narrative sequences; on the other hand, they assume the semantic elements [thematic roles] of an attributive or a functional order, with which the text is woven.[3]

An actor adds to its thematic roles a process of *individuation*.

> At the level of discourse [the thematic role] is manifested on the one hand as a qualification, an attribute, of the actor, and on the other hand, this qualification is, from the semantic point of view, only a designation that subsumes a field of functions [...]. Consequently, the minimal semantic content of a *role* is the same as that of an actor, with the exception that it does not contain the *seme of individuation*: a role is an animated figurative entity, but anonymous and *social*; an actor however is an *individual* incorporating and assuming one or several roles.[4]

---

1 Greimas, 1973b: 171.
2 Chabrol, 1973: 10.
3 Courtès, 1976: 95.
4 Greimas, 1970: 256.

This actorial individuation is perhaps the most important condition of the 'readability' of a narrative. Being individuated, an actor possesses a global invariant identity and a localized spatio-temporal existence. Assuming different actantial and thematic roles as the narrative progresses, it basically represents a place of transformation. Let us also observe that the actorial structure can be 'objectivized' (when each actor takes over a single actantial role) or 'subjectivized' (when several different actantial roles are syncretized in a single actor) as is the case, for instance, with the internal conflicts of 'psychological' heroes.

3.8.3 *Figurativization*

Figurativization is the most 'superficial' level. It endows values and themes with figures. Iconicity gives figures their characteristic forms, and produces a *referential illusion*, that is to say, the 'reality' effect of meaning. One of its essential components is the onomastic one responsible for anthroponyms, toponyms, chrononyms, etc.

3.8.4 *Spatio-temporal programing*

The discursive subcomponent called spatio-temporal programing depends upon the individuation of actors. Temporal programing converts the axis of presuppositions (that is to say, the logical order of narrative programs) into the axis of consecutions (a causal order of events). Temporal localization segments the narrative. And aspectualization

> transforms the narrative functions (of the logical type) into processes evaluated by an observer-actant situated within the discourse.[1]

Spatial programming is

> the procedure which consists in organizing the syntagmatic concatenation of partial spaces resulting from the spatial localization of the narrative programs.[2]

---

1 See, Greimas-Courtès, 1979: 387–388.

2 Ibid.: 295.

Now space is generally articulated and subdivided into partial spaces *conforming to the actantial organization.*

> Spatial localization must select first a space of reference – a zero space – from which the other partial spaces can be located [...]. This space of reference is called the *topic space,* while the surrounding spaces [...] are called *heterotopic spaces.* A further division of the topic space is often necessary, enabling to distinguish the *utopic* space [...] the place of performances (which in the mythic narratives is frequently underground, underwater or celestial), and the *paratopic* spaces where the competences are obtained.[1]

It seems clear that spatio-temporal localization belongs to the discursive syntax. However, we think it has also to be partially projected onto the deep structure. Greimas does it to a certain extent when he says:

> At the anthropomorphic level, the schemas [$d_1/{\sim}d_1$ and $d_2/{\sim}d_2$] correspond to the *isotopic spaces* which are places where the performances unfold. [...] Each space is constituted of two deixes which are *conjoined* (because they correspond to the same contradiction axis), but are *not consistent.* [...] On the other hand, the hypotactic axes ${\sim}d_2 \rightarrow d_1$ and ${\sim}d_1 \rightarrow d_2$ constitute heterotopic spaces whose deixes are *disjoined,* since they do not belong to the same schemas, but are *consistent,* since they are joined by the relation of presupposition.[2]

But he is thinking of deixes as places differentiated, articulated, and distributed by the semiotic square, without assuming any theory for the categorization of spaces. However, just as there exists an actantial structure underlying the actorial structure, *there exists a geometrico-topological infrastructure of spatio-temporal localization which is with respect to the latter what the actants are to the actors.* The *immanent actantial* (and no longer actorial) *spatiality* which we introduce here has nothing to do with the 'objective' space-time. It is an abstract (immanent and ideal) spatiality which becomes embedded in space-time by spatio-temporal localization. This localist spatiality is the key to the comprehension of actantial syntax as a deep topological syntax of operations. Just as individuation of actors and figurativity mask deep semio-narrative structures, so the objective space-time masks

---

1 Ibid.: 216.

2 Greimas, 1970: 176–77. The notations are as in Sec. 3.4.4.

the deep topological organization of actantiality. But at the deep level, space and time share a *syntactic* function.

3.9 *The narrative schema*

To conclude this panorama of semio-narrative theory, let us recall briefly the conception of the *global* structure of narratives (myths and tales), also referred to as the (canonical) *narrative schema*. The acquisition of competences (qualifying tests) open onto the performancial sequence (decisive test) which is a polemical confrontation of two narrative pathways. The performance is evaluated (axiologically) during the glorifying test by the final Sender-Awarder, 'the guardian of the contracts, of the equity of human relations, and of the truth about things and beings'.[1] Globally, a narrative depends upon an initial fiduciary contract. It is a device for liquidating an initial lack and restituting a transcendent order imperiled by a disequilibrium :

> Everything happens as if the narrative organization obeyed a principle of equilibrium which transcends and rules the human actions performed by the subjects.[2]

Through a series of contractual obligations, conflicts, agreements, ruptures, and reconciliations, it manifests the narrative schema as a 'formal pattern where happens to be inscribed 'the sense of life''[3] and as a

> hypothetical model of the general organization of narratives, that seeks to account for the forms used by the subject to conceive of his life as a project, realization, and destiny.[4]

And Greimas' profound conclusion is that:

> the very succession of tests, interpreted as a reversed order of logical presuppositions, seems to be governed by an intentionality that can be recognized *a*

1 Greimas, 1976: 24.
2 Ibid.: 22.
3 Ibid.: 22.
4 Greimas-Courtès, 1979: 245.

> *posteriori*, and which is similar to that used in genetics for explaining the development of an organism.[1]

This covert reference to biology is to be taken seriously: Greimasian theory is a *morphogenetic* theory of narratives. Following Propp, it analyzes actantial functions which are authentic narrative 'chreodes' (see Sec. I.1). Then it recognizes in the narrative schema global properties of equilibration and regulation. And finally, it reclaims the idea of an epigenetic 'landscape' governing the unfolding of the narrative. Therefore its mathematical schematization depends upon *a general mathematics of morphogenesis.*

4 Paul Ricoeur's observations

Greimasian theory involves a generative description of the various devices for the production of meaning. Its empirical and theoretical value is sure even if not universal. As T. Pavel observed,

> the introduction of the level of narrative structures would retain all its importance and its theoretical relevance if, instead of being a universal and perhaps trivial feature of the signifying objects, the presence/absence of this level could allow us to maintain a grand distinction between the narrative and the non-narrative semantic phenomena.[2]

In our panorama, we have focussed on the problem of conversion because its elucidation is of great consequence to the entire project of formalization. We have emphasized the fact that, owing to Greimas' 'logical' conception of the fundamental grammar, it becomes parodoxical. That is why we would like to discuss now some valuable observations of Paul Ricoeur who developed – from very different bases – a similar reflection.[3]

---

1 Ibid.: 371.

2 Pavel, 1980: 21.

3 Ricoeur, 1980. In this section, the page numbers of Ricoeur's analysis will be referred to in the text.

We will not discuss Ricoeur's assertion that if the Greimasian 'calculus' were to be always authenticated, 'there would be no happening, no surprise, nothing to tell (p. 10).' First of all, the operations and transformations 'predictable and computable' are nevertheless open to all the imaginable variations with respect to the canonical narrative schema. Though they constrain innovation, they do not obliterate it any more than grammatical rules obliterate the richness of language or prosodic rules obliterate poetic inventiveness. Further, innovation is basically situated at the discursive-figurative level, and the theory of semio-narrative structures describes only the authors' narrative *competence*. Finally, Ricoeur's objection has also to do with the semantic/semiologic vicious circle and can be answered by introducing the notion of semiotic pregnances (see Sect. 3.2.3).

We will take up Ricoeur's critical analysis of the conversion of fundamental logical operations into anthropomorphic syntactic doings.

> We will pose three questions. The first concerns the very principle of the distinction between the fundamental grammar and the surface narrative grammar. The second concerns the logical consistency of the constitutional model. The third concerns its 'narrativization'. (pp. 8–9)

The main criticism that Ricoeur directs against Greimas touches upon the relations between semiotics and linguistics.

> I do not question the right to base semiotics on linguistics. I question its articulation *before* linguistics. In this way semiotics and linguistics precede each other: the former by its generality, the latter by its specificity. The objection is not small as regards narrativity. If, indeed, semiotics and linguistics precede each other from different perspectives, then it is possible that semiotic analysis carried out within the context of an existing narrative organization, can rightfully construct *a priori* the semiotic square (or the semiotic squares) which structure(s) the text. In such a case, the semiotic analysis would possess a true heuristic capacity and will instruct truly how to read the text. But it can also be the case that semiotic analysis is feigned, I mean [...] less constructed *a priori* than reconstructed afterwards to satisfy the rules of semiotic organization. (pp. 8–9)

Ricoeur is denouncing here the methodological vicious circle that equates the deep immanence with a metalinguistic level reconstructed from manifestation. This is not a mere epistemological issue.

Its main purpose is to focus on the *logico-praxic* ambiguity of narrative syntax. At each step of the generative path, Greimas introduces an overcoming which he reformulates afterwards as an *equivalence* between two levels. This 'ambiguous character of the reduction of narrative to the logical or an overcoming of the logical in the narrative (p. 12)'[1] has much relevance to the eidetic transformation of taxonomic relations into syntactic operations (preconversion).

> As we will see, this question is posed at every level: is not the finality of an operation in the succeeding operation, and finally in the acheived idea of narrativity? And if the taxonomic model has been constructed in view of the syntactic operations grafted to it, is it not true that these operations in their turn become the conditions of narrativity only retrospectively, starting from their application in the surface narrative grammar – therefore in combination with features which appear only with specifications characteristic of the surface grammar? (p. 11)

This double postulation and this double requirement:

> on the one hand, extending in a progressive manner the logical strength of the initial taxonomic model to every level of narrativization, in such a way as to raise semiotics to the rank of a deductive science; on the other hand, constituting by a regressive process the scale of narrativity conditions from the final term, that is to say, the acheived idea of narrativity (p. 12)

becomes particularly critical in the conversion which syntagmatically converts the fundamental logical operations into a *generic* anthropomorphic doing

> which will ensure the transition towards doing in general, the nucleus of all anthropomorphic significations of the narrative. (p. 11)

There are two difficulties here. On the one hand, the reduction of the doing *in general* to a pure *syntactic* doing, and on the other hand the relationship between the logical and the praxical.

With regard to the first, Ricoeur notices a paralogism. The equivalence between the fundamental and the anthropomorphic syntax 'is ensured by the notion of syntactic doing, homogeneous both to the syntactic operations and to the ordinary doing' (p. 16).

---

1 Our emphasis.

But, 'substituting every verb of action with doing does not amount to transforming them into a syntactic doing' (p. 34).

Ricoeur raises here the problem of the possible equivalence between, on the one hand, the syntactic doing 'which reformulates the syntactic operations into anthropomorphic language' (p. 16) and, on the other hand, the generic doing which is 'the formal term substituting all the verbs of action' (p. 15). For Greimas, the syntactic doing is identical to the generic doing and this conversion bridges the gap between the logical and the praxic. For Ricoeur, such a suture is impossible.

Let us insist on this identity/difference between the syntactic doing and the generic doing. By refining the ordinary doing one gets a pure generic doing. By refining semantics one gets pure logico-conceptual operations. And the anthropomorphic syntactic doing converts one into the other. In other words, within Greimasian theory, *the anthropomorphic doing mediates between a logico-conceptual pole and an event-based praxic one.* In section 3, we have indicated the key idea which can resolve this difficulty. It consists in:

(i) conceiving the actantial syntax as a case-based localist syntax, refining the ordinary doing into a generic doing catastrophically schematized;

(ii) conceiving the fundamental semantic taxonomy as an articulation of semiotic pregnances, also catastrophically schematized;

(iii) interpreting the preconversion as a double reading (semantic/syntactic) of the *same* catastrophic schemas.

We agree with Paul Ricoeur when he insists that there can be no equivalence between the logical and the praxic as long as the semiotic and the linguistic, the semantic and the actantial share relationships of *mutual* precedence.

> The semiotic square brings its network of interdefined terms and its system of contradiction, contrariness and presupposition. The semantics of action brings the main significations of doing as well as the specific structures of statements referring to action. In this way, the surface grammar is a blended semiotico-praxic grammar. (p. 15)

'The discontinuity introduced by doing and its syntax between the logical and anthropological planes' (p. 16) is particularly clear if we take into consideration the fact that the syntactic subjects of doing are intentional actants (see Sec. 3.4.2 and 3.4.5).

> Nowhere the specificity of the semantics of action is more evident than in the passage from the statements of doing to the statements of being-able-to-do. On what basis can we say that wanting-to-do leads to the subsequent doing? Nothing in the semiotic square answers this question. (p. 15)

'The implicit phenomenology of a semantics of action' is *implied* by the notions of 'desire for realization' and 'realization of desire.'

> Once again, the significations brought by the semantics of action precede the semiotic square, even if the latter, by its logical simplicity, precede the complexity of the categories of the surface grammar. (pp. 15–16)

In another series of remarks, Ricoeur directs his criticism against another key notion, namely, performance, that is to say the polemical relations of confrontation and struggle which are 'the true anthropomorphic equivalents of the relations of the semiotic square (p. 17).' According to him, performance is 'the most characteristically blended – logical and praxic – unit of the entire narrative order (p. 19)' for confrontation and struggle are the pivotal figures of the semantics of action.

The first remarks concern the treatment of conflict as anthropomorphization of contradiction, though it refers to contrariness (see Sec. 3.4.4 §4 (b)). Indeed, Greimas assumes that the *S* / ~*S* conflict manifests the contrariness between the two schemas of contradiction of the semiotic square 'but, the weakening of the logical model is the only way possible to equate confrontation with contrariness as well as with contradiction. (p. 19)'[1]. The absence of an anthropomorphic representation of contrariness entails that the conflict cannot be associated with a *primitive* relation of the semiotic square (Problem 9). Whence the problem posed by the interpretation of negation as domination (Sec. 3.4.4 §4 (c)) and of assertion as acquisition of the value-object (Sec. 3.4.4 §4 (d)). The polemical relation of conflict is of a praxic nature. It reveals a sort of 'dialectical' negativity and anthro-

1 We may say that if the S/~S conflict is interpreted as the correspondence between two schemas then *S* and ~*S* no longer correspond to terms of the square, but to metaterms.

pomorphizes the *qualitative* opposition. It should therefore be treated as a *dynamical elementary* relation. But this is not logically possible because every formal logic excludes negativity and reduces it to a mere negation. As Ricoeur emphasizes:

> as far as the properly praxic relations of a polemical character are alien to the logical representation of contradiction – indeed even contrariness –, the construction of the semiotic square risks being reduced to an artifact of representation used by the semiotician for checking the correctness of his models. (p. 20)

That would be prejudicial for semiotics since

> the polemical structure of narration is what allows us to extend the initial *paradigmatic* articulation of the taxonomic model to every step in the *syntagmatic* progression of the narrative. (p. 34)

The second series of Ricoeur's remarks on performance concern the circulation of value-objects and the topological syntax of deixis-actants (see Sec. 3.4.5). As we have seen, this topological syntax enables us to identify the asymmetry of the conflict $S/{\sim}S$ with an axiological orientation of the contrariness between the two schemas. Since the terms of the schemas are substituted by the predicates of conjunction/disjunction we get the following reading of the semiotic square:

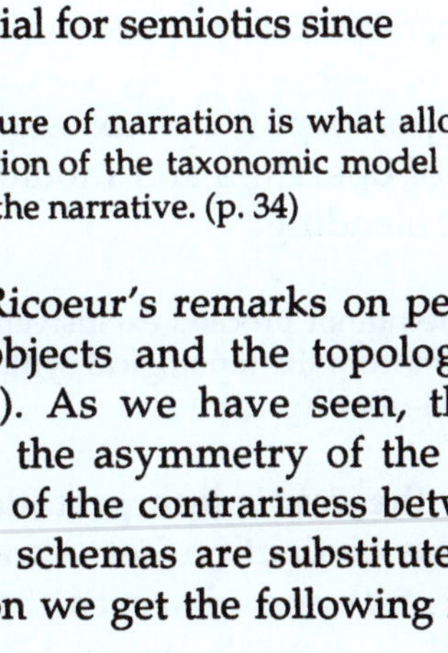

It allows us to interpret assertion as acquisition, and to account for the equivalence between an appropriation on the subject side and a correlative dispossession on the anti-subject side. The topological syntax of the deixis-actants thus governs (see 3.4.5):

(i) the forms of exchange, since the transfers lose their polemical quality and become identified with communicative transfers, and

(ii) the establishment of the senders and receivers of these transfers as virtual subjects of doing.

However, as Ricoeur remarks,

> how can one not notice that depriving and giving have meaning beyond just disjoining and conjoining? [...] What the last stage in the constitution of the model introduces is a phenomenology of acting-and being-acted-upon, within which notions like deprivation and gift take on a meaning. (p. 22)

Thus, while the anthropomorphic syntax is a logico-praxic blending, the topological syntax of operators and modal values is, according to Ricoeur, a logico-pathic blending.

> The operational value cannot proceed exclusively from the logical aspects of attribution, but in turn from the topological syntax and from the semantics of acting and being-acted-upon. (p. 23)

Ricoeur then turns to the *deictic* interpretation of the terms of the semiotic square (for the distinction between (iso)topic and heterotopic spaces, see Sec. 3.8.4) and observes that since the correlation between schemas involves a weak, or even analogical, contrariness, the 'concordance' of the disjoint deixes weakens the hypotactic relation of implication. According to him, it is indeed this weakening which explains how the predictable and computable syntactic operations can also *create* values. In order to make this possible, 'logicality must in some ways be inadequate to the creativity of the narrative' (p. 24). The inadequacy exists as long as the concordance of deixes is not governed by logical implication.

> [Deixes] conform to the Aristotelian notion of *dianoia* associated with the *mythos* of the narrative. [...] It is the history of culture that generates the schematism of these *dianoia* and *mythoi,* which is a matrix of relations and operations of weak logical character. (p. 24)

In a final remark, Ricoeur takes up the question of the contractual and adjudicative sequences framing the narrative (the global narrative schema). He asks why they are not included in the surface grammar even when they correspond to the hypotactic relations of the taxonomical model.

We see that Ricoeur's critical evaluation of the Greimasian analytical apparatus refers to the links between the paradigmatic and syntagmatic axes.

> Greimas' topological considerations represent [...] the most extreme attempt to stretch the paradigmatic as far as possible into the core of the syntagmatic. (p. 27)

For Greimas, each syntagmatic addition must have a paradigmatic equivalent describable by a conversion procedure.

> If we consider narration in its syntagmatic perspective where each narrative programme appears as a process consisting of gains and losses of values, improvements and impoverishments of the subjects, then one notices that each step on the syntagmatic axis correspond to (and is defined by) a topological displacement on the paradigmatic axis.[1]

For Ricoeur, on the contrary, there exists at every stage of the generative path, a real syntagmatic *supplement* which cannot be reduced to a conversion.

> The fundamental question that Greimas' attempt poses is that of the cumulative and progressive generation of levels of depth of the semiotic model. Does the device of increasing levels of depth have the function of extending at each new stage the initial qualities of the taxonomic model? Or, on the contrary, is it this introduction, at each stage, of new syntactico-semantic components (e.g., anthropomorphic representation, addition of figurativiy) that confers on the device its fecundity? (p. 26)

> If it is true, as we have tried to show, that some syntagmatic innovation appears at each level, at first under the pressure of a semantics of action, and later with the praxic-pathic categories of polemics and exchange, *then the power for innovation belongs to these praxic-pathic investments and not to the initial taxonomic model.* (p. 26) [2]

---

1 Greimas, 1976b: 25.
2 Our emphasis.

Ricoeur insists on the *blended* nature of the Greimasian model, for 'without appropriate additions of a clearly syntagmatic type, [...] the taxonomic model would remain static and sterile' (p. 27).

We see that this critique is centered around three main arguments:

(i) For principled eidetic reasons, the polemical negativity is derivable neither from taxonomic relations of contradiction nor from the syntactic operation of negation.

(ii) There are syntagmatic supplements which cannot be obtained by conversion from the fundamental grammar.

(iii) The praxic-pathic dimension

> brings into play a semantics of action, which in turn brings into play a syntax, whose intelligibility is itself blended: phenomeno-logical and linguistic. (p. 17)

From our own perspective we will make the following observations.

(i) For Ricoeur's critique to be really valid, it would be necessary to formalize the phenomenology of action so that it could be included in the formal model.

(ii) Since polemical negativity can be morphodynamically formalized (catastrophe theory is a general dynamical theory of conflicts), a morphodynamical reformulation of the fundamental grammar should be developed.

(iii) The hypothesis of a syntagmatic irreducible supplement to the paradigmatic dimension is valid only if we think of conversion as a mere metalanguistic *equivalence*. Thus, a deeper understanding of conversion is essential.

(iv) In the catastrophist schematism, conversion *is not* a mere equivalence but a double reading of specific dynamical structures.

In fact, the morphodynamical schematism integrates a modeling of the phenomenology of action. This cognitively embedded phenomenology represents the *etic* counterpart of Greimas' emic structural syntax.

5 Schematization of the undefinables

Our sketch of structural semiotics has led us to the conclusion that for formalizing the theory of semio-narrative structures, we must have a very precise type of mathematics that can account for:

(i) semiotic 'prégnances' and their diffusion;

(ii) articulation and categorization of substrata;

(iii) topologico-actantial syntax conceived from a localist perspective;

(iv) intentionality;

(v) the eidetic of conflict;

(vi) preconversion.

As we will show in another work,[#] Morphodynamics can do the job. But to apply this sort of mathematics to semiotics and to confer it a function of *objective determination* is absolutely not the same thing as to elaborate a conceptual-descriptive theory. It presupposes the latter, but is intended to convert it from 'meta-physics' to 'physics'. That is why, by way of conclusion, we shall make some observations on epistemology.

5.1 *Greimasian epistemology*

As we have seen, the Greimasian theory is structural (Saussurian-Hjelmslevian) and actantial-conceptual [1] (Tesnierian-Fillmorian). It depends on a principle of 'ontological independence of the semiotic form'[2] conceived of as *Gestalt*. In this sense, Greimas' 'logicism' is not a true one. Greimas himself often contrasted it with the formalist conceptions of language. Narrative syntax is neither syntagmatic and derivational (as generative grammars) nor categorial (as categorial

---

\# Petitot [1992].

1 'Conceptual' in the cognitive sense of 'conceptual structure'.

2 Greimas-Courtès, 1979: 155.

grammars of Ajdukiewicz, Bar-Hillel or Montague). It is an actantial-conceptual syntax for which syntactic relations are *meaningful.*[1]

> *Formal syntaxes* are elaborated without any reference to signification. [...] On the contrary, *conceptual syntaxes* consider the syntactic relations as *signifying* (since they belong to the form of the content), even if they are abstract in nature, and can be considered as logical relations. For semiotics this is a fundamental choice.[2]

Greimasian theory was developed in a well-defined epistemological framework based on a specific conception of formalization and metalanguage. Greimas resorts to Hjelmslev's idea as per which the semiotic metalanguage *is itself a semiotics,* 'that is, a hierarchy – not of words or phrases – but of definitions involving either the semiotic system or its process'.[3]

This hierarchical and definitional conception of metalanguage has considerable consequences with respect to the question of formalization. Let us present here some details of these issues which have been partially dealt with in Sec. I.3.1 and in Chapter II.

1. Because of its definitional nature, the hierarchical construction ultimately takes recourse to primitive *undefinable* concepts which 'can be considered as hypothetical universals.'[4]

2. Since the metalanguage is a language of description, formalization will consist, first of all in providing a *formal expression* to the undefinable primitive concepts.

> Following the tradition of Saussure and Hjelmslev, where signification is the creation and/or the apprehension of 'differences', [our theory] will include all those concepts which, while undefinable, are necessary for establishing the definition of the elementary structure of signification. This conceptual explicitation leads to a *formal expression* for the selected concepts: considering the structure as a relational network, we will have to construct an axiomatic se-

1 See Sec. I.3.3. In this sense, narrative actantial grammar is akin to cognitive grammars (Talmy, Langacker, Lakoff, Jackendoff, Fauconnier, etc.).
2 Ibid.: 378.
3 Ibid.: 225.
4 Ibid.: 345.

> miotic which will be essentially in the form of a typology of relations. (presupposition, contradiction, etc.).[1]

3. Once the axiomatic semiotics has been elaborated, then a deductive framework will 'produce linguistics as formal language, that is, as 'pure algebra'.'[2]

4. The elaboration of a 'minimal formal language' takes us to the question of 'the selection [...] of the *systems of representations* within which [the semiotic theory] can formulate its methods and models'.[3]

Now, for Greimas, these languages and models 'are only different ways of representing the same phenomena and the same 'reality''.[4] In other words, one assumes a phenomenal 'reality' *independent* of the formalization and describable by means of concepts derived from the undefinable primitives. Within such an epistemology, formal objects like the semiotic square or the constituent-structure trees of generative grammars are mere graphic representations of the structural organization. Greimas thought he could rule out that way the philosophical problem of the ontological status of structures. Viewed as a 'relational network whose nodes constitute the terms, [...] endowed with a specific internal organization', a structure is an 'autonomous entity'

> whose ontological status need not be questioned and which, on the contrary should be bracketed so as to render the concept operational.[5]

5.2 *The undefinables as universals*

The concept of structure is noematic and the undefinable primitives needed for its formalization are *transcendental categories* operating as *universals* of language.[6] Let us cite the major ones.

---

1 Ibid.
2 Ibid.: 225.
3 Ibid.: 345.
4 Ibid.: 225.
5 Ibid.: 361.
6 Ibid.

1. The concept of *discontinuity* :

> Being undefinable, the category *continuous/discontinuous* is to be retained in the epistemological inventory of the undefined concepts. [...] It is often said that the projection of the *discontinuous* onto the continuous is the first condition for making the world intelligible. The problematic of this 'projection' is a matter of general epistemology, and is not proper to semiotics. [...] In semiotics, every entity # is to be considered as continuous prior to its articulation.[1]

2. The concept of *relation*:

> [A relation] establishes concomitantly both the identity and the difference of two or more entities. [...] Such a definition is however only an interdefinition articulating semiotic universals, since the terms of identity and difference require as part of their own definition, the undefinable concept of relation.[2]

Relation is further developed in the concepts of *difference, opposition,* and *junction.*

3. The concept of *term,* correlative to that of relation.

4. The concepts of *identity* and *difference.*

5. The concept of *distinction* that establishes differences: '*Distinction* is an undefined concept to be retained in the epistemological inventory'.[3]

6. The concept of *condition* that establishes the concepts of *presupposition, reciprocal presupposition,* and *implication.*

7. The concept of *totality*: 'In semiotics, totality can be envisaged first as an undefinable concept belonging to the epistemological inventory of universals'.[4]

'Entity' translates the French word 'grandeur'. 'Magnitude' would be inappropriate for it is too quantitative.

1 Greimas-Courtès, 1979: 101.

2 Ibid.: 314.

3 Ibid.: 109.

4 Ibid.: 397.

8. The concept of *discreteness*, which is a sub-category of the category of *totality*, and is contrary to those of *globality* and *indivisibility*: 'As an undefined concept, *discreteness* is to be retained in the epistemological inventory of the undefinables'.[1]

9. The concepts of *assertion* and *negation* with which we can define *contradiction*:

> The relation of *contradiction* is the relation which exists between the two terms of the binary category *assertion/negation*. Given that the notions 'relations', 'term', 'assertion', and 'negation' refer to undefinable concepts, the definition is situated at the deepest and the most abstract level of semiotic articulation.[2]

10. The concepts of *transformation* and *operation*: the basic transformations are assertion and negation; an operation is the effective aspect of a transformation.

11. The concepts of *subject* and *object* which allow us to define a *junction* as a subject/object relation.

12. The concept of *description*: according to Hjelmslev, description 'is the best example of an undefinable concept'.[3]

13. The concepts of *expression* and *content*, and of course the concept of *meaning*: 'A property common to all semiotics, the concept of *meaning* is undefinable'.[4]

5.3 *The foundational aporia of the form of meaning*

Greimas' epistemology is a direct consequence of the fact that its object is the *form of meaning*, and that meaning cannot be objectivized. Meaning is not a phenomenon amenable to sense experience. It is ungraspable as such. This 'foundational aporia' pervades the theory

---

1 Ibid.: 67–68.
2 Ibid.
3 Ibid.: 92.
4 Ibid.: 348.

and forces it to be conceptual-descriptive, metalinguistic and constructed on the basis of undefinables.

> Today semantics seems to have ruled out the apprehensions of many linguists expressed in the famous Bloomfieldian position as per which though meaning exists, one has no means to study it scientifically. [...] It required a revolution in thinking – replacing the convictions regarding the description of linguistic 'facts' with the idea that linguistics is primarily a theoretical construction that seeks to account for phenomena that are otherwise ungraspable – for semantics to be accepted and recognized as a constructed language capable of describing the object-language.[1]

The foundational aporia of the form of meaning has been well explained by Greimas in *Du Sens*. Meaning is an immediate consciousness manifested only by its form. But this form appears only as transformations of meaning.

> The production of meaning takes place only as the transformation of the given meaning; consequently, the production of meaning has to be viewed as a meaningful formation, irrespective of the contents to be transformed. *Meaning, as form of meaning, can be defined as the possibility of transforming meaning.*[2]

> Only such a semiotics of forms [of meaning] can provide in the foreseeable future the language for describing meaning. For, indeed, *the form of meaning is nothing but the meaning of meaning.*[3]

Hence the vicious circle between the semiotic forms of meaning and a semiotics of the forms of meaning. It is an aspect of the hermeneutic circle and of the aporia of metalanguage:

> Every metalanguage that can be thought of for describing meaning is not only a signifying language, but it is also nominalizing, that is, it freezes all intentional dynamism into a conceptual terminology.[4]

In order to overcome this paradox we have to raise the semiotic forms of meaning to the status of *phenomena* and constitute their own specific *objectivity*.

---

1 Greimas-Courtès, 1979: 326 (our emphasis).
2 Greimas, 1970: 156 (our emphasis); see also Sec. 3.2.3.
3 Ibid.: 17.
4 Ibid.: 8.

5.4 *The 'dead core' of the theory and the necessity of schematization*

As a conceptual, formal, and descriptive metalinguistic construction based on undefinable primitives, the Greimasian theory comes up against an epistemological obstacle very similar to that identified by T. Pavel with respect to Chomsky:

> It is not possible to include all the grammatical, categorial or functional notions in the derivation and one is constrained to define some of these notions from primitive notions. Now, this solution poses insuperable problems. As Putnam already noted in 1960, by selecting some notions as primitives, Chomsky avoids defining them. [...] These categories are in fact the 'dead core' of the theory.[1]

We would like to counter this conception of a theory as a hierarchical metalinguistic construction with a truly scientific one. The key issue here is the *mathematical* schematization of the undefinable primitives, which deeply transforms the gnoseological status of semiotic theory. When the primitive categories and universals are schematized, they acquire a mathematical content which strongly *constrains* the construction of derived concepts. The languages of representation that has been hitherto considered as mere graphical visualizations now yields true falsifiable models.

We see that the deepest problem is to elaborate *a non formalist (non symbolic) theory of form*. What Greimas needed was certainly a 'pure algebra' of semiotic forms, but this 'algebra' was not formal in the logical sense.

> While we can define logic as the form of the content used for verifying the *linguistic formulations of the scientific form of the universe as expression* (this scientific form is called 'semantics' by logicians), the logic that we need for semantics is a kind of algebra of the form of linguistic expression, with which we could verify the articulations of the semantic structure.[2]

Such a 'logic' must be qualitative and morphological. It depends upon *the general mathematics of morphologies and morphogenesis*. As we have seen in this chapter, there exists a deep analogy between Greimasian semiotics and the classical concepts of biomorphology

1 Pavel, 1980: 19.
2 Greimas, 1970: 43.

(especially embryogenesis): the taxonomic model as differentiation of substrata, the Proppian functions as narrative 'chreodes', the narrative schema as 'epigenetic landscape,' etc. They point to a deep problem : *in the expression 'algebra of forms' it is the term 'form' that conditions the term 'algebra' and not vice versa.* In other words *it is the mathematical content assigned to the primitive 'form' that determines the 'algebraic' structure of the universe of forms.* In the symbolic perspective, the undefinable primitive 'form' is assigned a purely symbolic content, and composition, iteration, and combinatorial complexification are formalized in terms of symbols. In the morphodynamical perspective, they are formalized in dynamical terms.

The mathematical models adequate and appropriate for semiotics should be:

(i) mathematical models of morphogenesis;

(ii) models that incorporate within their structure a procedure of phenomenological reduction and of phenomenological 'abduction';

(iii) models sharing the same degree of formality and universality as logical symbolizations.

As morphodynamical models satisfy these three conditions, we are justified in proposing their application to the semiotic region. As models of differentiation of substrata, of articulation of substances, of conflict of regimes, of colocalisation of positional proto-actants, they yield a dynamical geometry of the 'discontinuous' and 'positional' primitives. That is why there exists an equivalence between the schematization of structural categories and the application of morphodynamical models to structuralism.

We will show in another work # that they allow us to solve the theoretical problems of semiotics sketched in this chapter.

See Petitot, 1992.

CONCLUSION

1. This investigation of a 'physics' of meaning was developed within the framework of a general morphological and structural eidetics. As regards phonetics, we have shown that morphodynamical models enable us to resolve the 'foundational aporia' between the *substance*-based and the *form*-based conceptions of the phonological code. As regards structural syntax, we have shown that the morphodynamical shematization of the localistic hypothesis enables us to *deduce* case universals, and to *configurationally* define their content. While examining the main domains of the structural field, we have validated the hypothesis that Morphodynamics is able to solve the central theoretical problem of structuralism described in Chapter I.

That Morphodynamics can have such a relation of objective determination with structuralism is in general not well understood.[#] But, let us refer again to Krzysztof Pomian who, in his article 'Structure' in the *Enciclopedia Einaudi* insists on the following points :[1]

(i) one of the most important scientific developments in the twentieth century is the division of human sciences into theoretical sciences and historical sciences;

(ii) from the theoretical perspective, objects are real though non-empirical and they exert a function of objective determination on the empirical facts;

(iii) though structures are not phenomena in the strict sense of the term, structuralism ensures their ontological promotion;

(iv) structuralism oscillates between a Jakobsonian 'nomogenesis' inherited from phenomenology and a Hjelmslevian formalism inspired by the Vienna Circle logicism;

(v) by introducing an abstract mathematical theory of morphogenesis into the structural region, the dynamical structuralism

---

\# The situation is now completely reversed, due to the fantastic developments of neuromimetic dynamical models which implement morphodynamical structures in neural nets.

1 Pomian, 1981. See also Pomian, 1977.

of Catastrophe Theory *synthesizes* Jakobson's and Hjelmslev's programs, and integrate them to the *natural* sciences.

At the end of this work, we hope we are able to assert that this is one of the major aspects of Morphodynamics.

2. We may be perhaps reproached for having focussed with some extent on philosophical questions. This was however necessary for the two following reasons.

(i) As structures are not phenomena in the strict sense of the term and as modeling can be relevant only for phenomena, a structural theory can take on only three forms. It may be conceived (as in the formalist view) as a meta-semiotics for which modeling is reduced to a mere logical symbolization and/or graphic representation. Or, it may be reduced (as in the eliminativist psychological view) to an analysis of natural psychical processes. Or – and this was our mathematical-phenomenological position – it 'roots' structures in the mind while still claiming that they share emergent (supervenient) objective features. Now, any hypothesis concerning emerging autonomous levels of reality raises deep philosophical problems.

(ii) Let us call *mathematical constitution* the assignment of a specific mathematical content to the regional categories of a certain domain of reality. The philosophical legitimation of a mathematical constitution is a *transcendental* problem. Any objectivity must be *constituted,* that is founded on the legitimation of a mathematical constitution by a transcendental one. Our purpose was to apply this 'constitutive' perspective (traditional in physics) to structuralism and, in particular, to the semiotic realm. This entailed first a phenomenalization and a naturalization, next an objectivization, and finally a mathematization of meaning, three research programs which requires also to tackle many philosophical difficulties. The possibility of objectivizing semiotic categories is far from being evident since many philosophers think that, on the contrary, we have to 'semiotize' the categories of objectivity.

3. At the 1981 Albi Conference *Le Savoir et le Croire* (*Knowing and Believing*),[1] Paul Ricoeur presented what he called the 'memory of the problems' that links semiotics with the philosophical genealogy. He emphasized three 'reductions' of ontology that followed one another during the modern period:

(i) the Kantian reduction to phenomena;

(ii) the phenomenological reduction to lived experience;

(iii) the semiotic reduction to discourse.

Each of these reductions completely reorganized the philosophical heritage.

Our aim is to reformulate the semiotic reduction as a reduction to a new type of cognitive phenomena. For this, we need a new type of 'transcendental aesthetics' which allows us to select a specific type of mathematics for schematizing and constructing mathematically the structural categories (in much the same way as in physics the geometry of space-time allows the schematization-construction of physical categories). Our thesis was that the localistic concept of position and the morphological concept of discontinuity were adapted to such a task. They are in some sense the 'pure intuitions' of structural objectivity. It is a geometry of positions and a dynamics of discontinuities which allow us to objectivize the form of meaning and to constitute what we call a *Physics of Meaning*.

---

1 Albi, 1983.

BIBLIOGRAPHY

Abbs, J. H., Sussman, H. M., 1971, Neurophysiological feature detectors and speech perception: a discussion of theoretical implications, *Journal of Speech and Hearing Research, 14, I,* 23–36.

Abraham, R., 1972, *Introduction to Morphology,* Department of Mathematics, 9, University of Lyon I.

Albi, 1983, *On Believing* (H. Parret ed.), New York, Berlin, de Gruyter.

Almansi, G. (ed.), 1970, Structuralism, *20th Century Studies,* 3.

Anderson, J. M., 1971, *The Grammar of Case, Towards a Localistic Theory,* Cambridge University Press.

Anderson, J. M., 1975*a,* La grammaire casuelle, *Langages, 38,* Paris, Didier-Larousse, 18–64.

Anderson, J. M., 1975*b,* Maximi Planudis in Memoriam, *Langages, 38,* Paris, Didier-Larousse, 81–103.

APP, 1983, *L'auto-organisation. De la physique au politique* (P. Dumouchel et J.-P. Dupuy eds.), Cerisy Conference, Paris, Le Seuil.

Bach, E., 1965, Linguistique structurelle et philosophie des sciences, *Diogène, 51.*

Bachmann, L. (ed.), 1828, *Anecdota Graeca,* II, Leipzig, Georg Olms, Hildesheim, 1965.

Badcok, C. R., 1975, *Lévi-Strauss : structuralism and sociological theory,* London, Hutchinson.

Barthes, R., 1966, Introduction à l'analyse structurale des récits, *Communications, 8.*

Bastide, R. (ed.), 1962, *Sens et usage du terme structure dans les sciences humaines et sociales,* Paris, La Haye, Mouton.

Benoist, J. M., 1975, *La révolution structurale,* Paris, Grasset & Fasquelle.

Benveniste, E., 1966, *Problèmes de linguistique générale,* Paris, Gallimard.

BGRS, 1977, *Le Bulletin du Groupe de Recherches sémio-linguistiques, 1,* Ecole des Hautes Etudes en Sciences Sociales, Paris.

BGRS, 1981, *Le Bulletin du Groupe de Recherches sémio-linguistiques, 17*, Ecole des Hautes Etudes en Sciences Sociales, Paris.

BGRS, 1982, *Le Bulletin du Groupe de Recherches sémio-linguistiques, 24*, Ecole des Hautes Etudes en Sciences Sociales, Paris.

Boudon, R., 1968, *A quoi sert la notion de 'structure' ?*, Paris, Gallimard.

Boudon, R., 1973, Structure dans les sciences humaines, *Encyclopaedia Universalis, 15*, 438–440.

Boudon, P., 1981, Les ordres de la figuration, *Communications, 34*, Paris, Le Seuil.

Brandt, P. A., 1976, Sémiotique : sémantique et symbolique, *SES*, 1976, 144–172.

Brandt, P. A., 1982*a*, Quelques remarques sur la véridiction, *Actes sémiotiques, IV*, Ecole des Hautes Etudes en Sciences Sociales, 5–19.

Brandt, P. A., 1982*b*, Noise et narrativité, *Actes sémiotiques, V*, 24, Paris, Ecole des Hautes Etudes en Sciences Sociales, 46–55.

Brandt, P. A., 1992, *La charpente modale du sens. Pour une sémio-linguistique morphogénétique et dynamique*, Aarhus University Press & Amsterdam, John Benjamins.

Broekman, J. M., 1974, *Structuralism : Moscow-Prague-Paris*, Boston, Reidel.

Brown, R., 1973, *A First Language*, Cambridge, Harvard University Press.

Bryant, J. S., 1978, Feature Detection Process in Speech Perception, *Journal of Exp. Psych.* (Human Perception and Performance), 4, 4, 610–620.

BTC, *Bibliography on Catastrophe Theory*, Mathematics Institute, University of Warwick.

Buffon, 1774, *Œuvres complètes*, t. IV, Paris, Imprimerie Royale.

Cassirer, E., 1945, Structuralism in Modern Linguistics, *Word, 1*, 2.

Cassirer, E., 1923, *Das Erkenntnisproblem in der Philosophie und der Wissenschaft der neueren Zeit*, Berlin, Verlag Bruno Cassirer.

Chabrol, C., 1973, De quelques problèmes de grammaire narrative et textuelle, *Sémiotique narrative et textuelle* (C. Chabrol ed.), Paris, Larousse.

Chenciner, A., 1973, Travaux de Thom et Mather sur la stabilité topologique, *Bourbaki Seminar*, 424.

Chenciner, A., 1980, Singularités des fonctions différentiables, *Encyclopaedia Universalis.*

Chenciner, A., 1984, Systèmes dynamiques différentiables, *Encyclopaedia Universalis*, Paris.

Chomsky, N., 1965, *Aspects of the Theory of Syntax*, Cambridge, Mass., MIT Press.

Chomsky, N., 1966, *Cartesian Linguistics*, New York, Harper & Row.

Chomsky, N., 1968, *Language and Mind*, New York, Harcourt, Brace & World.

Comrie, B., 1974, Causatives and Universal Grammar, *Transactions of the Philological Society*.

Comrie, B., 1977, In Defense of Spontaneous Demotion: The Impersonal Passive, *GR*, 1977, 47–58.

Coquet, J. C., 1979, *Prolégomènes à l'analyse modale: le sujet énoncant*, Documents de recherche du Groupe de Recherches Sémio-linguistiques, 3, Ecole des Hautes Etudes en Sciences Sociales, Paris.

Coquet, J. C., 1982, *Sémiotique. L'école de Paris*, Paris, Hachette.

Courtès, J., 1973, *Lévi-Strauss et les contraintes de la pensée mythique*, Paris, Ed. Mame.

Courtès, J., 1976, *Introduction à la sémiotique narrative et discurisive*, Paris, Hachette.

Culioli, A., 1970, La formalisation en linguistique, *Considérations théoriques à propos du traitement formel du langage*, Documents de Linguistique quantitative, 7, Dunod.

Culioli, A., Desclés, J.-P., 1981, *Systèmes de représentation linguistiques et métalinguistiques*, Formal Linguistics Laboratory, University of Paris 7.

Cutting, J. E., Rosner, B. S., 1974, Categories and Boundaries in Speech and Music, *Perception and Psychophysics, 16*, 3, 564–570.

Damisch, H., 1973, Structure et art, *Encyclopaedia Univasalis, 15*, 440–442.

Danchin, A., 1977, Stabilisation fonctionnelle et épigenèse, *L'Identité* (C. Lévi-Strauss' Seminar, J.-M. Benoist ed.), Paris, Grasset, 185–221.

Delattre, P., 1968, From Acoustic Cues to Distinctive Features, *Phonetica, 18*, 4, 198–230.

Delattre, P., 1971, *Système, structure, fonction, évolution : essai d'analyse épistémologique*, Paris, Doin.

Delattre, P. *et al.*, 1973, Structure et fonction, *Encyclopaedia Universalis, 15*, 442–445.

Deleuze, G., 1973, A quoi reconnaît-on le structuralisme?, *Histoire de la philosophie* (F. Chatelet ed.), Paris, Hachette.

Derrida, J., 1967, *L'écriture et la différence,* Paris, Le Seuil.

Diehl, R. L. *et al.*, 1978, Contrast Effects on Stop Consonant Identification, *Journal of Exp. Psych. (Human Perception and Performance)*, 4, 4, 599–609.

Ducrot, 0., 1968, *Le structuralisme en linguistique,* Paris, Le Seuil.

Eco, U., 1963, The Analysis of Structure, *The Critical Moment,* London, Faber.

Eco, U., 1968, La struttura assente, Milano, Bompiani.

Eco, U., 1976, *A Theory of Semiotics,* Bloomington, Indiana University Press.

Ehrmann, J. (ed.), 1966, Structuralism, *Yale French Studies.*

Eimas, P. D., 1974, Auditory and Linguistic Processing of Cues for Place of Articulation by Infants, *Perception and Psychophysics, 16,* 3, 513–521.

Eimas, P. D., 1980, Infant Speech Perception : Issues and Models, *Perspectives on Mental Representation* (M. Garrett, J. Melher, E. Walker ed.), Hillsdale, New Jersey, L. Erlbaum.

Eimas, P. D., Corbitt, J. D., 1973, Selective Adaptation of Linguistic Feature Detectors, *Cognitive Psychology,* 4, 99–109.

EJM, 1979, *Elaboration et justification des modèles* (P. Delattre et M. Thellier eds.), Paris, Maloine.

Fillmore, C., 1966, A Proposal Concerning English Prepositions, *Monograph Series on Languages and Linguistics, 19,* 19–33.

Fillmore, C., 1968, The Case for case, *Universals in Linguistic Theory* (E. Bach, R. T. Harms, eds.), 1–88, New York, Holt, Rinehart & Winston.

Fillmore, C., 1969, Toward a Modern Theory of Case, *Modern Studies in English,* (A. D. Reibel, S. A. Schane eds.), Englewood Cliffs, New Jersey, Prentice-Hall, 361–375.

Fillmore, C., 1970, The Grammar of Hitting and Breaking, *Readings in English Transformational Grammar* (A. R. Jacobs, R. S. Rosenbaum eds.), Waltham, Massachusetts, 120–133.

Fillmore, C., 1971*a*, Types of Lexical Information, *Semantics* (D. D. Steinberg, L. A. Jacobovitz eds.), London, New York, Cambridge University Press, 370–392.

Fillmore, C., 1971*b*, Some Problems for Case Grammar, *Monograph Series on Languages and Linguistics*, 24. See also *Langages*, 38, Paris, Didier-Larousse.

Fillmore, C., 1971*c*, Verbs of Judging, *Studies in Linguistic Semantics* (C. Fillmore, D. T. Langendoen eds.), New York, Holt, Rinehart & Winston, 273–289.

Fillmore, C., 1972, Subjects, Speakers and Roles, *Semantics of Natural Languages* (D. Davidson, G. Harman eds.), Dordrecht, Reidel.

Fillmore, C., 1975, Quelques problèmes posés à la grammaire casuelle, *Langages*, 38, 65 80.

Fillmore, C., 1977, The Case for case reoppened, *Syntax and Semantics*, 8, *Grammatical Relations* (P. Cole, J. M. Sadock eds.), Academic Press, 59–81.

Fink, S. R., 1978, Case Grammar and Valence Theory at a Stalemate? Their Relevance for Semantic Memory, *VSG*, 1978, 177–190.

Gandillac, M. de, Goldmann, L., Piaget, J., 1965, *Entretiens sur les notions de genèse et de structure*, Paris, La Haye, Mouton.

Gil, F., 1981, Sistematica e classificazione, *Enciclopedia Einaudi, XII*, Torino, Einaudi, 1024–1044.

Gil, F., Petitot,J., 1981, Uno/Molti, *Enciclopedia Einaudi, XI*, Torino, Einaudi.

Glucksmann, M., 1974, *Structuralist Analysis in Contemporary Social Thought*, Boston, Routledge & Kegan Paul.

Golubitsky, M., Guillemin, V., 1973, *Stable Mappings and their Singularities*, Graduate Texts in Mathematics, 14, New York, Heidelberg, Berlin, Springer.

Gould, S. J., 1977, *Ontogeny and Philogeny*, Cambridge, Harvard University Press.

Gould, S. J., Eldredge, N., 1977, Punctuated Equilibria: the Tempo and Mode of Evolution Reconsidered, *Paleobiology*, 3, 2, 115–151.

GR, 1977, *Grammatical Relations*, Syntax and Semantics, 8 (P. Cole, J. M. Sadock eds.), New York, Academic Press.

Greimas, A. J., 1966, *Sémantique structurale*, Paris, Larousse.

Greimas, A. J., 1970, *Du Sens*, Paris, Le Seuil.

Greimas, A. J., 1973*a*, Un problème de sémiotique narrative : les objets de valeur, *Sémiotiques textuelles* (M. Arrivé, J.-C. Coquet eds.), *Langage*, 31, Paris, Didier-Larousse.

Greimas, A. J., 1973*b*, Les actants, les acteurs et les figures, *Sémiotique narrative et textuelle* (C. Chabrol ed.), Paris, Larousse.

Greimas, A. J., 1976*a*, Talk with F. Nef, *SES*, 1976.

Greimas, A. J., 1976*b*, Preface to *Courtès*, 1976.

Greimas, A. J., 1983, *Du Sens, II*, Paris, Le Seuil.

Greimas, A. J., Courtès, J., 1979, *Sémiotique, Dictionnaire raisonné de la théorie du langage*, Paris, Hachette.

Gross, M., 1975, *Méthodes en syntaxe*, Paris, Hermann.

Guillaume, P., 1979, *La psychologie de la forme*, Paris, Flammarion.

Günther, H., 1978, Valence in Categorical Syntax, VSG, 1978, 127–156.

Harris, Z. S., 1951, *Structural Linguistics*, University of Chicago Press.

Harris, Z. S., 1970, *Papers in Structural and Transformational Linguistics*, New York, Reidel.

Hawkes, J., 1977, *Structuralism and Semiotics*, London, Methuen.

Hénault, A., 1979, *Les enjeux de la sémiotique*, Paris, Presses Universitaires de France.

Hénault, A., 1983, *Narratologie, Sémiotique générale*, Paris, Presses Universitaires de France.

Hjelmslev, L., 1935, *La catégorie des cas*, München, Wilhelm Fink Verlag, 1972.

Hjelmslev, L., 1968, *Prolégomènes à une théorie du langage*, Paris, Minuit.

Hjelmslev, L., 1971, *Essais linguistiques*, Paris, Minuit.

Husserl, E., 1913, *Ideen zu einer reinen Phänomenologie udn phänomenologischen Philosophie, I*, Husserliana 3, (W. Biemel ed.), The Hague, M. Nijhoff, 1950.

Jacob, S. W., Francone, C. A., 1970, *Structure and Function in Man*, London, Saunders.

Jakobson, R., 1971, *Selected Writings*, Paris, La Haye, Mouton.

Jakobson, R., Fant, G., Halle, M., 1952/1967, *Preliminaries to Speech Analysis*, Cambridge, MIT Press.

Jakobson, R., Lévi-Strauss, C., 1962 & 1972, 'Les chats' de Charles Baudelaire, *L'Homme;* Charles Baudelaire's 'Les chats', *The structuralists from Marx to Lévi-Strauss* (de George ed.), Garden City, Doubleday.

Jakobson, R., Waugh, L., 1980, *La charpente phonique du langage* (trad. A. Kihm), Paris, Minuit.

Johnson, D. E., 1974*a*, *Toward a Theory of Relationnaly-based Grammar*, Thesis, University of Illinois, Urbana.

Johnson, D. E., 1974*b*, *On Relationnal Constraints on Grammars*, IBM, Thomas J. Watson Research Center, Yorktown Heights, New York.

Johnson, D. E., 1974c, On the Role of Grammatical Relations in Linguistic theory, *Papers from the Tenth Regional Meeting of the Chicago Linguistic Society*, University of Chicago.

Johnson, D. E., 1977, On Relationnal Constraints on Grammar, *GR*, 1977, 151–178.

Johnson, D. E., *Ergativity in Universal Grammar*, IBM, Thomas J. Watson Research Center, Yorktown Heights, New York.

Johnson, D. E., 1979, *Toward a Theory of Relationnally Based Grammar*, New-York, Garland Publishing.

Johnson, D. E., Postal, P., 1980, *Arc Pair Grammar*, Princeton, Princeton University Press.

Jusczyk, P. W., 1980, Is there a Phonetic Basis for Speech Perception During Infancy?, *Perspectives on Mental Representation* (M. Garrett, J. Mehler, E. Walker eds.), Hillsdale, New York, L. Erlbaum.

Kant, I., 1790. *Kritik der Urtheilskraft*, Kants gesammelte Schriften, Band V, Preussische Akademie der Wissenschaften, Berlin, Georg Reimer, 1913.

Katz, J. J., Fodor, J. A. (eds.), 1964, *The Structure of Language*, Prentice-Hall, Englewood Cliffs.

Keenan, E. L., 1972, On Semantically Based Grammar, *Linguistic Inquiry*, 3, 413–462.

Keenan, E. L., 1975, Somme Universals of Passive in Relationnal Grammar, *Papers from the Eleventh Regional Meeting of the Chicago Linguistic Society* (R. E. Grossman, L. J. Sam, T. J. Vance eds.), University of Chicago.

Keenan, E. L., 1976, Towards a Universal Definition of 'Subject', *Subject and Topic* (C. Li ed.), New York, Academic Press.

Keenan, E. L., Comrie, B., 1972, Noun-Phrase Accessibility and Universal Grammar, *Congress of the Linguistic Society of America*.

Ladefoged, P., 1962, *Elements of Acoustic Phonetics*, Chicago, University of Chicago Press.

Ladefoged, P., 1972, Phonetic Prerequisites for a Distinctive Feature Theory, *Papers in Linguistics and Phonetics to the Memory of Pierre Delattre* (A. Valdman ed.), Berlin, New York, Mouton.

Laughlin, C. D., 1974, *Biogenetic Structuralism*, Columbia University Press.

Leach, E., 1976, *Culture and Communication*, Cambridge University Press.

Leibniz, 1706, Letter to des Bosses, *in* C. Fremont, *L'être et la relation*, Paris, Vrin, 1981.

Lehiste, I. (ed.), 1967, *Readings in Acoustic Phonetics*, Cambridge, MIT Press.

Lévi-Strauss, C., 1949, *Les structures élémentaires de la parenté*, Paris, Presses Universitaires de France.

Lévi-Strauss, C., 1958, *Anthropologie structurale*, Paris, Plon.

Lévi-Strauss, C., 1964–1971, *Mythologiques: Le Cru et le Cuit; Du Miel aux Cendres; L'Origine des Manières de table; L'Homme nu*, Paris, Plon.

Libéra, A. de, 1976, Note on Arild Utaker's 'On binary opposition', *SES*, 1976, 49–55.

Liberman, A. M., Cooper, F. S., Shankweiler, D. P., Studdert-Kennedy, M., 1967, Perception of the Speech Code, *Psychological Review*, *74*, 6, 431–461.

Lindblom, B., 1972, Some Phonetic Null Hypotheses for a Biological Theory of Language, *Proceedings of the Ninth International Congress of Phonetic Sciences*, Vol. 2, Copenhague, 33–40.

Lisker, L., Abramson, A. S., 1964, A Cross Language Study of Voicing in Initial Stops: Acoustic Measurement, *Word*, *20*, 384–422.

Lisker, L., Abramson, A. S., 1970, Discriminability along the Voicing Continuum.: Cross Language Tests, *Proceedings of the Sixth International Congress of Phonetic Sciences*, Prague.

Lu, Y. C., 1976, *Singularity Theory and an Introduction to Catastrophe Theory*, Berlin, New York, Springer.

Luria, A. R., 1975, Scientific Perspectives and Philosophical Dead Ends in Modern Linguistics, *Cognition*, *3*, 4, 377.

Macksey, R., Donato, E. (ed.), 1970, *The Languages of Criticism and the Sciences of Man. The Structuralist Controversy*, Baltimore, Johns Hopkins Press.

Mac Neilage, P. F., 1979, Speech Production, *Proceedings of the Ninth International Congress of Phonetic Sciences*, *1*, Copenhague, 11–39.

Malmberg, B., 1974, *Manuel de phonétique générale*, Paris, Picard.

Manjali, F., 1991, *Nuclear Semantics*, New Dehli, Bahri.

Maranda, P. (ed.), 1966, *Structural Analysù of Oral Tradition*, University of Pennsylvania.

Marin, L., 1971, *Sémiotique de la passion*, Paris, Aubier-Montaigne.

Massaro, D. W., 1972, Issues of Speech Perception, *Proceedings of the Ninth International Congress of Phonetic Sciences*, 2, Copenhague, 474–481.

Mehler, J., Bertoncini, J., 1980, Infant's Perception of Speech and other Acoustic Stimuli, *Infant Behavior and Development*, 2.

Miller, J. L., 1975, Properties of Feature Detectors for Speech : Evidence from the Effects of Selective Adaptation on Dichotic Listening, *Perception and Psychophysics*, *18*, 6, 389–397.

Miller, J. D., *et al.*, 1976, Discrimination and Labelling of Noise-buzz Sequences with Varying Noise-lead Times : an Example of Categorical Perception, *Jour. Acoust. Soc. Am.*, *60*, 410–417.

Miller, G. A., Jonhson-Laird, Ph. N., 1976, *Language and Perception*, Cambridge, Harvard University Press.

Nef, F., 1979, Case Grammar VS Actantial Grammar : some Remarks on Semantic Roles, *Text VS Sentence* (J. S. Petöfi ed.), Hamburg, Helmut Buske, 634–653.

Olmsted Gary, J., Keenan, E. L., 1977, On Collapsing Grammatical Relations in Universal Grammar, *GR*, 1977, 83–120.

Osgood, C. E., 1971, Where Do Sentences Come From?, *Semantics* (D. Steinberg, L. A. Jacobovitz eds.), Cambridge University Press.

Ouellet, P., 1982, *Le sens de forme du sens: Essai sur l'ordre du Logos*, Thesis, Paris, University of Paris 7.

Pastore, R. E. *et al.*, 1977, Common-factor Model of Categorical Perception, *Journal of Experimental Psychology* (*Human Perception and Performance*), 3, 4, 686–696.

Pavel, T., 1979, Phonology in Myth-Analysis, *Text VS Sentence* (J. S. Petöfi ed.), Hamburg, Helmut Buske Verlag, 654–666.

Pavel, T., 1980, *Modèles génératifs en linguistique et en sémiotique*, Documents de Recherche du Groupe de Recherches Sémio-linguistiques, *20*, Ecole des Hautes Etudes en Sciences Sociales, Paris.

Perlmutter, D., (ed.) 1983/1984, *Studies in Relationnal Grammars*, I/II, Chicago, University of Chicago Press.

Petitot, J., 1977*a*, Introduction à la théorie des catastrophes, *Mathématiques et Sciences humaines, 59.*

Petitot, J., 1977*b*, Topologie du carré sémiotique, *Etudes littéraires*, Québec, University of Laval.

Petitot, J., 1977c, Identité et catastrophes, *L'identité* (C. Lévi-Strauss' Seminar, J.-M. Benoist, ed.), Paris, Grasset.

Petitot, J., 1978, Catastrophes, *Encyclopaedia Universalis.*

Petitot,J., 1979*a*, Infinitesimale, *Enciclopedia Einaudi, VII*, Torino, Einaudi, 443–521.

Petitot,J., 1979*b*, Locale/Globale, *Enciclopedia Einaudi, VIII*, Torino, Einaudi, 429–490.

Petitot, J., 1979*c*, Hypothèse localiste et théorie des catastrophes, note sur le débat, *TLTA*, 1979, 516–524.

Petitot, J., 1979*d*, Saint-Georges: Remarques sur l'espace pictural, *Sémiotique de l'espace*, Paris, Denoël-Gonthier.

Petitot, J., 1982*a*, *Pour un schématisme de la structure : de quelques implications sémiotiques de la théorie des catastrophes*, Thesis, Ecole des Hautes Etudes en Sciences Sociales, Paris.

Petitot, J., 1982*b*, *Paradigme catastrophique et perception catégorielle*, Documents du Centre d'Analyse et de Mathématique sociales, Ecole des Hautes Etudes en Sciences Sociales, Paris.

Petitot, J., 1982*c*, Sur la signification linguistique de la théorie des catastrophes, *Mathématiques et Sciences humaines, 79*, Ecole des Hautes Etudes en Sciences Sociales, Paris, 37–74.

Petitot, J., 1982*d*, A propos de la querelle du déterminisme : de la théorie des catastrophes à la Critique de la faculté de juger, *Traverses, 24*, 134–151.

Petitot, J., 1982*e*, Sur la décidabilité de la véridiction, *Actes sémiotiques, IV*, 31, Ecole des Hautes Etudes en Sciences Sociales, Paris, 21–40.

Petitot, J. (ed.), 1982*f*, *Logos et théorie des catastrophes*, René Thom's Cerisy Conference, Genève, Patiño, 1989.

Petitot, J. (ed.), 1982*g*, Aspects de la conversion, *Actes sémiotiques, V*, 24, Ecole des Hautes Etudes en Sciences Sociales, Paris.

Petitot, J., 1983*a*, Théorie des catastrophes et structures semio-narratives, *Actes sémiotiques, V*, 47–48, Ecole des Hautes Etudes en Sciences Sociales, Paris, 5–37.

Petitot, J., 1983*b*, *Structure*, Documents du Centre d'Analyse et de Mathématique sociales, Ecole des Hautes Etudes en Sciences Sociales, Paris. Eng. Trans. *Encyclopedic Dictionary of Semiotics* (Th. Sebeok, ed.), Vol. 2, 991–1022, Walter de Gruyter, New-York, 1986.

Petitot, J., 1985, *Les catastrophes de la parole. De Roman Jakobson à René Thom*, Paris, Maloine.

Petitot, J., 1992, *Physique du sens*, Paris, Editions du CNRS.

Petitot, J., Thom, R., 1983, Sémiotique et théorie des catastrophes, *Actes sémiotiques, V*, Ecole des Hautes Etudes en Sciences Sociales, Paris, 47–48.

Piaget, J., 1968, *Le structuralisme*, Paris, Presses Universitaires de France.

Pisoni, D. B., 1979, On the perception of specch sounds as biologically significant signals, *Brain, Behav. Evol., 16*, 330–350.

Pomian, K., 1977, Catastrofe, *Enciclopedia Einaudi, II*, Torino, Einaudi, 789–803.

Pomian, K., 1981, Struttura, *Enciclopedia Einaudi, XIII*, Torino, Einaudi, 723–764.

Postal, P. M., 1971, The Method of Universal Grammar, *Method and Theory in Linguistics* (P. Garvin ed.), La Haye, Mouton.

Postal, P. M., 1974*a*, *Report of Work Done as Part of the MSSB Workshop on Constraints on Grammar*, IBM Thomas J. Watson Research Center, Yorktown Heights, New York.

Postal, P. M., 1974*b*, *On Raising*, Cambridge, MIT Press.

Poston, T., Stewart, I., 1978, *Catastrophe Theory and its Applications*, Boston, Londres, Pitman.

Potts, T., 1978, Case-grammar as Componential Analysis, *VSG*, 1978, 399–457.

Prigogine, I., 1980, *Physique, temps et devenir*, Paris, Masson.

Propp, V., 1970, *Morphologie du conte* (trad. C. Ligny), Paris, Gallimard.

QS, 1973, *Qu'est-ce que le structuralisme?*, O. Ducrot, J. Todorov, D. Sperber, M. Safouan, F. Wahl, Paris, Le Seuil.

Raccani, R., Eco, U., 1969, *I sistemi di segni e lo strutturalismo sovietico*, Milano, Bompiani.

Renfrew, C., Cooke, K. L., 1979, *Transformations. Mathematical approaches to culture change*, New York, Academic Press.

Repp, B. H., *et al.*, 1978, Perceptual Integration of Acoustic Cues for Stop, Fricative, and Affricate Manner, *Journal of Experimental Psychology (Human Perception and Performance)*, 4, 4, 621–637.

Ricœur, P., 1980, *La grammaire narrative de Greimas*, Documents du Groupe de Recherches Sémio-linguistiques, 15, Ecole des Hautes Etudes en Sciences Sociales, Paris.

Riemann, B., 1854, *Über die Hypothesen, welche der Geometrie zu Grunde liegen*, Gesammelte Mathematische Werke, Berlin-Heidelberg, Springer, 1990.

Robey, D., 1973, *Structuralism, an Introduction*, Oxford, Clarendon Press.

Ross, J. R., 1974, There, there (there (there (there...))), *Papers from the Tenth Regional Meeting of the Chicago Linguistic Society*, University of Chicago.

Ruffié, J., 1982, *Traité du vivant*, Paris, Fayard.

Ruwet, N., 1967, *Introduction à la grammaire générative*, Paris, Plon.

Saunders, P. J., 1980, *An Introduction to Catastrophe Theory*, Cambridge, Cambridge University Press.

Saussure, F., 1915, *Cours de linguistique générale*, Paris, Payot.

Schlesinger, I. M., 1971, Production of Utterances and Language Acquisition, *The Ontogeneisù of Grammar* (D. I. Slobin ed.), New York, Academic Press, 63–101.

Sebeok, T. A., Osgood, Ch. (eds.), 1965, *Psycholinguistics*, Bloomington, Indiana University Press.

Segre, C. *et al.*, 1965, *Structuralismo e critica*, Milano, Saggiatore.

SES, 1976, *Structures élémentaires de la signification* (F. Nef ed.), Bruxelles, Ed. Complexe.

Siegel, J. A., Siegel, W., 1977, Categorical Perception of Tonal Intervals: Musicians can't tell *sharp* from flat, *Perception and Psychophysics*, *21*, 5, 399–407.

Smith, B. (ed.), 1982, *Parts and Moments, Studies in Logic and Formal Ontology*, Vienne, Philosophia Verlag, Analytica.

SSP, 1979, *Structural Stability in Physics* (W. Güttinger et H. Eikemeier eds.), Berlin, New York, Springer.

SSTC, 1976, *Structural Stability, the Theory of Catastrophes and Applications in the Sciences* (P. Hilton éd.), Lecture Notes in Mathematics, 525, Berlin, New York, Springer.

Starosta, S., 1973, The Faces of Case, *Language Sciences, 25*, 1–14.

Starosta, S., 1975, Les visages du cas, *Langages, 38*, 104–128.

Steiner, R., 1884, Preface to J. W. von Gœthe's *Die Metamorfose der Pflanzen*, Paris, Triades, 1975.

Stevens, K., 1972*a*, The Quantal Nature of Speech, *Human Communication, a Unified View* (P. B. Denes, E. E. David Jr. eds.).

Stevens, K. N., 1972*b*, Bases for Phonetic Universals in the Properties of the Speech Production and Perception Systems, *Proceedings of the Ninth International Congress of Phonetic Sciences*, vol. 2, Copenhague, 53–59.

Stevens, K. N., Blumstein, S. R., 1978, Invariant Cues for Place of Articulation in Stop Consonants, *Jour. Acoust. Soc. Am., 64*, 5, 1358–1368.

Studdert-Kennedy, M. *et al.*, 1970, Motor Theory of Speech Perception: a Reply to Lane's Critical Review, *Psychological Review, 77*, 3, 234–249.

Tesnière, L., 1959, *Eléments de syntaxe structurale*, Paris, Klincksieck.

Thom, R., 1966, Une théorie dynamique de la morphogenèse, *Towards a Theoretical Biology I* (C. H. Waddington ed.), University of Edinburg Press.

Thom, R., 1968*a*, Biologie et structuralisme, *Towards a Theoretical Biology*, Vol. III (C. H. Waddington, ed.), University of Edinburg Press.

Thom, R., 1968*b*, Topologie et signification, *L'Age de la Science, 4*, Dunod, 1–24 (see also Thom, 1980*a*).

Thom, R., 1969, Topological Models in Biology, *Topology, 8*, 313–335.

Thom, R., 1970, Topologie et Linguistique, *Essays on Topology and Related Topics* (G. de Rham's Festschrift), Springer, 226–248.

Thom, R., 1971, Le rôle de la topologie dans l'analyse sémantique, *Symposium de Sémantique*, Urbino (see also Thom, 1980*a*).

Thom, R., 1972*a*, *Stabilité structurelle et morphogenèse*, New York, Benjamin, Paris, Ediscience.

Thom, R., 1972*b*, Langage et catastrophes : Eléments pour une sémantique topologique, *Bahia Symposium on Dynamical Systems*, New York, Academic Press, 619–654.

Thom, R., 1973*a*, Un protée de la sémantique: L'information, *Unesco Conference*, Venise (see also Thom, 1980*a*).

Thom, R., 1973*b*, Sur la typologie des langues naturelles : essai d'interprétation psycho-linguistique, *L'analyse formelle des langues naturelles* (M. Gross, M. Halle, M. P. Schutzenberger eds.), Mouton, 233–248 (see also Thom, 1980*a*).

Thom, R., 1973*c*, De l'icône au symbole : esquisse d'une théorie générale du symbolisme, *Les Cahiers internationaux du symbolisme*, 22–23 (see also Thom, 1980*a*).

Thom, R., 1975, La théorie des catastrophes : état présent et perspectives, *Dynamical Systems*, Lecture Notes in Mathematics, New York, Heidelberg, Berlin, Springer, 468, 366–372 and 384–389.

Thom, R., 1978*a*, Morphogenèse et imaginaire, *Circé, 8–9*, Paris, Editions Lettres modernes.

Thom, R., 1978*b*, La double dimension de la grammaire universelle, in Thom, 1978*a*.

Thom, R., 1979a, Modélisation et scientificité, *Elaboration et justification des modèles* (P. Delattre, M. Thellier eds.), I, Paris, Maloine, 21–30.

Thom, R., 1979*b*, Le rationnel et l'intelligible, *Orbetello Conference*, Académie internationale de Philosophie des sciences.

Thom, R., 1980*a*, *Modéles mathématiques de la morphogenèse*, Paris, Christian Bourgois.

Thom, R., 1980*b*, *The Role of Mathematics in Present-day Science*, Hanovre, Union Internationale de Philosophie des Sciences.

Thom, R., 1980*c*, Prédication et grammaire universelle, *Fundamenta Scientiae, 1*, 24.

Thom, R., 1980*d*, *Parabole e Catastrofe* (G. Giorello ed.), Milano, Il Saggiatore, Flammarion, 1983.

Thom, R., 1981, Morphologie du sémiotique, *Semiotic Inquiry, 1*, 4, 301–309.

Thom, R., 1983, Structures cycliques en sémiologie, *Actes sémiotiques, V*, Ecole des Hautes Etudes en Sciences Sociales, Paris, 47–48, 38–58.

TLTA, 1979, *Théories du langage, théories de l'apprentissage, le débat Chomsky-Piaget*, Paris, Le Seuil.

Utaker, A., 1974, *Semiotic Square and Binary Opposition*, Paris, La Haye, Mouton.

Viet, J., 1965, *Les méthodes structuralistes dans les sciences sociales*, Paris, La Haye, Mouton.

VSG, 1978, *Valence, Semantic Case and Grammatical Relations* (W. Abraham ed.), Studies in Language Companion Series A, Amsterdam, Benjamin.

Waddington, C. H., 1956, *Principles of Embryology*, London, Allen & Unwin.

Waddington, C. H., 1957, *The Strategy of Genes*, London, Allen & Unwin.

Waddington, C. H., 1965–1972, *Towards a Theoretical Biology*, Edinburgh University Press.

Webster, G., Goodwin, B., 1981, History and Structure in Biology, *Perspectives in Biology and Medecine*, 39–69.

Webster, G., Goodwin, B., 1982, The origin of species: a structuralist approach, *J. Sociol. Biol. Struct.*, *5*, 15–47.

Wildgen, W., 1981, Archetypal Dynamics in Word Semantics: an Application of Catastrophe Theory, *Words, Worlds and Contexts* (H. J. Eikmeyer, H. Reiser eds.), New York, Walter de Gruyter, 234–296.

Wildgen, W., 1982, *Catastrophe Theoretic Semantics. An Elaboration and Application of René Thom's Theory*, Amsterdam, Benjamin.

Wildgen, W., 1999, *De la grammaire au discours. Une approche morphodynamique*, Bern, Peter Lang.

Willems, D., 1978, A la recherche d'une grammaire des cas. Les rapports avec la syntaxe et le lexique, *VSG*, 1978, 243–260.

Zeeman, C., 1977, *Catastrophe Theory*, Addison-Wesley, Mass.

European Semiotics / Sémiotique Européenne

European Semiotics has been created to publish research in the field of Dynamic Semiotics. This is a discipline which has its origins in Phenomenology, Gestalt Theory and Structuralism, and looks for inspiration to the work of René Thom. The central problem of meaning is viewed primarily from the perspective of Linguistics but the theoretical approaches of Perception and Cognitive Studies are also employed. Questions of Aesthetics and the Philosophy of Language are also included in its scope.

The series originates in an initiative launched by a group of researchers in Semiotics in Denmark, Germany, Spain, France and Italy, who have been meeting regularly for several years. The continuing work of this group lends credibility to the concept of a European Semiotics.

The series will publish monographs, collected papers and conference proceedings of a high scholarly standard.

La collection *Sémiotique Européenne* se propose de publier les recherches en cours dans le cadre de la sémiotique dynamique. Cette discipline est héritière de la phénoménologie, de la gestaltthéorie et du structuralisme. Elle a trouvé dans l'œuvre de René Thom une nouvelle source d'inspiration et elle théorise la question de la signification non seulement à partir de la linguistique, mais aussi de la théorie de la perception et de la cognition. Elle relève aussi des domaines de l'esthétique et de la philosophie du langage.

Cette collection est initiée par un groupe de chercheurs européens provenant d'Allemagne, du Danemark, d'Espagne, de France et d'Italie. Ce groupe se réunit depuis plusieurs années et démontre qu'une sémiotique européenne peut exister.

La série comprend des monographies, des recueils d'articles et des actes de colloque d'un niveau scientifique élevé.

In der Reihe *Europäische Semiotik* werden in erster Linie Forschungsergebnisse aus dem Bereich der dynamischen Semiotik publiziert. Diese Disziplin hat das theoretische Erbe der Phänomenologie, der Gestalttheorie und des Strukturalismus angetreten und benutzt das Werk von René Thom als Quelle der Inspiration. Das zentrale Problem der Bedeutung wird nicht nur aus der Perspektive der Linguistik, sondern auch aus der Sicht der Wahrnehmungs- und Kognitionstheorie angegangen. Außerdem gehören Fragestellungen der Ästhetik und der Sprachphilosophie zum thematischen Bereich.

Die Publikationsreihe geht auf die Initiative einer Gruppe von Semiotikern aus Dänemark, Deutschland, Spanien, Frankreich und Italien zurück, die sich seit einigen Jahren regelmäßig trifft. Ihre Existenz ist der Beweis dafür, daß eine europäische Semiotik möglich ist. Die Reihe bietet Raum für Monographien, Sammelbände sowie Tagungsberichte auf hohem wissenschaftlichen Niveau.

Volume 1 Wolfgang Wildgen: De la grammaire au discours
Une approche morphodynamique
xi + 351 pp. 1999. ISBN 3-906762-03-3 / US-ISBN 0-8204-4226-7

Volume 2 Lene Fogsgaard: Esquemas copulativos de SER y ESTAR
Ensayo de semiolingüística
382 pp. 2000. ISBN 3-906764-22-2

Volume 3 Jean Petitot: Morphogenesis of Meaning
xii + 279 pp. 2003. ISBN 3-03910-104-8 / US-ISBN 0-8204-6858-4

Volume 4 Per Aage Brandt: Spaces, Domains, and Meaning
Essays in Cognitive Semiotics
ISBN 3-03910-227-3 / US-ISBN 0-8204-7006-6. In Vorbereitung.